Topics in Safety, Risk, Reliability, and Quality

Volume 23

Series Editor

Adrian V. Gheorghe

For further volumes:
http://www.springer.com/series/6653

Kumar Neeraj Jha

Determinants of Construction Project Success in India

Springer

Kumar Neeraj Jha
Department of Civil Engineering
Indian Institute of Technology Delhi
New Delhi
India

ISSN 1566-0443
ISBN 978-94-007-6255-8 ISBN 978-94-007-6256-5 (eBook)
DOI 10.1007/978-94-007-6256-5
Springer Dordrecht Heidelberg New York London

Library of Congress Control Number: 2013932648

Printed on acid-free paper

Springer is part of Springer Science+Business Media (www.springer.com)

Preface

It gives me immense pleasure in presenting this book to you. Construction industry in most of the countries is infamous for schedule and cost overrun, poor quality, large number of disputes, and many other ills. This is truer in case of developing countries such as India. Not many researches have been conducted to understand the reasons behind poor performances in a scientific manner. Even the rare successes achieved in some of the projects implementation are not well documented.

This book has been a result of research conducted at the IIT Delhi as part of my Ph.D. work and my subsequent supervisions of Ph.D.s and Masters Dissertations. The research systematically attempts to find out the critical success and failure attributes/factors across the four performance criteria: schedule, cost, quality, and dispute. It also shows which success factor(s) is (are) more relevant at a given project performance level. The book also presents the success traits for the success of a project. It reinforces the importance of commitment, coordination, and competence (3Cs) in achieving the desired project performance. The readers will also find ways in which a project's performance can be predicted. The need of a project coordinator is felt increasingly in large modern projects. The book also presents the required traits of a project coordinator.

The research is based on an empirical setting and sound research methodology. It utilizes various appropriate techniques, such as factor analysis, multinomial logistic regression, structural equation modeling, and neural network suitable for achieving a particular objective. Further most of the materials presented in the book are peer reviewed.

Construction industry contributes a lot in a country's GDP and is a prime employment generator. Growth in construction propels growth in many other industries. It is imperative to achieve good performance in this industry by implementing projects successfully. The book is an attempt to understand the performance enablers and retarders.

Each chapter begins with an abstract and is then organized into well-defined sections and subsections. Each chapter is summarized and concluded and at the end the relevant references are provided. Summary of relevant literature is also cited in each chapter. The book will be useful to different stakeholders of construction. It would also prove to be helpful to research students of different streams, I am eager to receive the comments from the readers of the book.

Kumar Neeraj Jha

Acknowledgments

A book of such nature would not have been possible without the support and assistance of a number of people. I would like to thank my supervisor Prof. K. C. Iyer and my students in particular Dr. S. Z. S. Tabish and Mr. C. T. Chockalingam. I am thankful to the publishers mentioned below for granting me permission to produce from my earlier works published by them.

S. No	Title	Journal Name	Publisher	Volume, Issue, and Year
(1)	Critical Factors Affecting Schedule Performance: Evidence from Indian Construction Projects	Journal of Construction Engineering and Management	American Society of Civil Engineers	132(8), 2006
(2)	Success Traits for a Construction Project	Journal of Construction Engineering and Management	American Society of Civil Engineers	138(10), 2012
(3)	Commitment, coordination, competence and the iron triangle	International Journal of Project Management	Elsevier	25(5), 2007
(4)	Critical determinants of project coordination	International Journal of Project Management	Elsevier	24(4), 2006
(5)	What attributes should a project coordinator possess?	Construction Management & Economics	Taylor & Francis	24(9), 2006
(6)	Prediction of schedule performance of Indian construction projects using an artificial neural network	Construction Management & Economics	Taylor & Francis	29(9), 2011

(continued)

(continued)

S. No	Title	Journal Name	Publisher	Volume, Issue, and Year
(7)	Ranking and classification of construction coordination activities in Indian projects	Construction Management & Economics	Taylor & Francis	25(4), 2007
(8)	Critical Factors Affecting Quality Performance in Construction Projects	Total Quality Management & Business Excellence	Taylor & Francis	17(9), 2006

I am thankful to Nathalie Jacobs and Cynthia Feenstra of Springer in guiding and supporting me throughout the review and publication process. I am also thankful to the typesetting team for their hard work. I am thankful to my research scholars Dilip, Amit, and Manish for the help extended in proof reading process. My sincere thanks to Prof. Anupam Dewan for all the help extended during the publication process. I am thankful to my colleagues at the Department of Civil Engineering—Indian Institute of Technology Delhi.

I am thankful to my friend and philosopher Prof. Lalit Manral—Assistant Professor of Strategic Management at University of Central Oklahoma for inspiring me to take up the academic profession. I am always motivated by his maverick approach in different walks of life. No words can suitably express my gratitude toward my parents for what they have done all through these years. Last but not the least; I am thankful to Arti (wife), Srijan (elder son) and Sajal (younger son) for their continued support.

Kumar Neeraj Jha

Contents

Chapter 1
Introduction

Abstract Chapter begins with a brief description of a construction project, its various phases and their importance. A brief review of literature on project success and the criteria employed to measure the success is presented. The performances of construction projects in general are not up to the mark and thus it is imperative to understand the factors for ensuring success in a project. The reasons why a project fails are also equally important. It would be also interesting to predict the performance of an ongoing project. These are some of the reasons why this study has been undertaken. Past knowledge points out the role of coordination in ensuring project success, however, what constitutes project coordination and what traits should be possessed by a project coordinator also need to be explored and accordingly they have been considered as one of the objectives of the study. Chapter ends with a brief description of contents of different chapters.

1.1 A Construction Project

Construction is an act or a process of constructing. It consists of a series of actions to produce either a new set of buildings and infrastructure or may involve alterations in the existing buildings and infrastructure (Radosavljevic and Bennett 2012). A construction project is a part of construction work that is being attempted or undertaken. A project involves a series of complex or interrelated activities and tasks that consume resources to achieve some specific objectives. It has to be completed within a set of specifications under a limited budget (Munns and Bjeirmi 1996; Pinto and Slevin 1988a).

Construction projects involve varying manpower and their duration can range from few weeks to more than five years. However, in some cases, the duration may be very long as in case of the Sardar Sarovar Project of India which took almost 60 years to become operational.

K. N. Jha, *Determinants of Construction Project Success in India*,
Topics in Safety, Risk, Reliability, and Quality 23, DOI: 10.1007/978-94-007-6256-5_1,

Although the act or process stages involved may be similar in different projects, yet each project is 'unique' and 'temporary' in nature and so is its management. Here the term 'unique' means that every project is different in some distinguishing way from all other projects and the term 'temporary' means that every project has a definite beginning and an end (PMBOK 2000).

For many organizations, projects are a means to respond to those requests that cannot be addressed within the organization's normal operational limits. A project may involve a single unit of one organization or may extend across organizational boundaries, as in case of joint ventures and partnering. A project is regarded as a key to accomplish the business strategy of any organization, as it is the means by which strategy is implemented (PMBOK 2000) and a project is therefore not an isolated event but is a realization of objectives through concerted efforts of different participants/stakeholders in various phases of the project life cycle.

1.2 Phases of a Construction Project

A construction project passes through different distinct phases or stages during its life cycle. It is also common to have some overlaps between the characteristics of two phases. The phases are described in a number of life cycle models. Each life cycle model has some specificity that can explain a particular aspect of a project. For example, *straightforward project life-cycle* approach advocates two very basic phases: a 'pre-project phase' consisting of identifying possible projects called project concepts and the 'implementation phase' of project. *Control-oriented project life-cycle*model suggests three main stages of a construction project: planning, execution, and operation (Bonnal et al. 2002). Munns and Bjeirmi (1996) adopt six stage model of the life of a project as given below.

1. Conception phase—the idea for project is birthed within the client organization and its feasibility determined.
2. Planning phase—the method to achieve the original idea is planned and designed.
3. Production phase—the plans are converted into physical reality.
4. Handover phase—the finished project is handed over to the client for use.
5. Utilization phase—the client makes use of the finished project.
6. Closedown phase—the project is dismantled and disposed of at the end of its useful life.

While developing the framework to assess project success of design/build projects Chan et al. (2002) have used three phases of a project, viz., pre-construction phase, construction phase, and post construction phase. On the other hand, Pinto and Slevin (1988b) have considered the four-phases of a project life cycle, i.e., the conceptual, planning, execution, and termination phases for finding the relative importance of critical success factors. The characterization of phases according to above models is shown schematically in Fig 1.1.

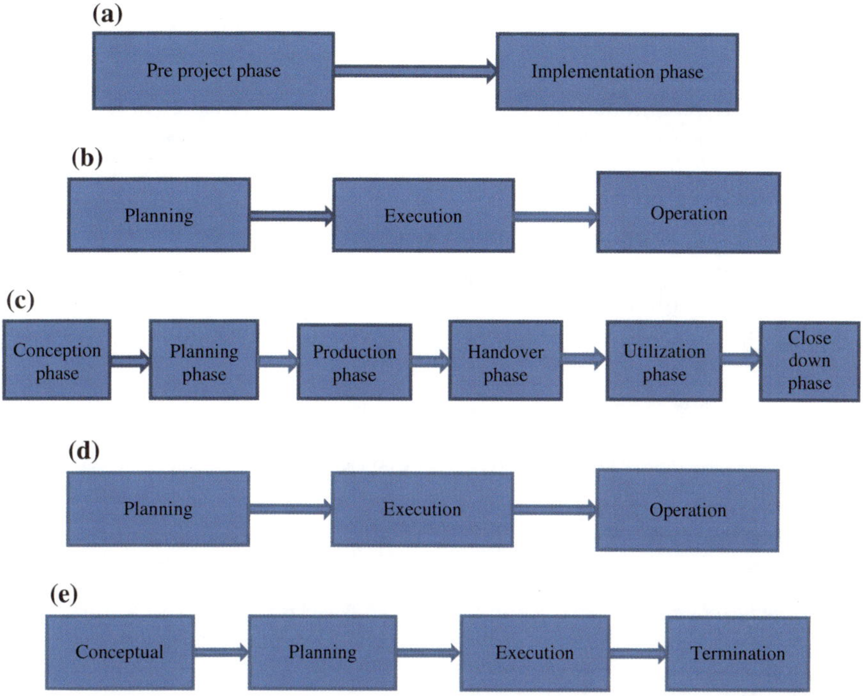

Fig. 1.1 Project phase characterization in different models. **a** Project phases in straightforward project life-cycle model. **b** Project phases in Control-oriented project life-cycle model. **c** Six stage model adopted by Munns and Bjeirmi (1996). **d** Three phase model adopted by Chan et al. (2002). **e** Four-phases of a project life cycle adopted by Pinto and Slevin (1988b)

From the project management literature, it is also observed that the phases play a dominant role in several decisions making. For example, Pinto and Slevin (1989) find that phases of a construction project decide the relative importance to be assigned to various project performance attributes in its life cycle. According to de Wit (1988), phases of a construction project govern the importance of various project success criteria. Spitz (1982) concludes that change in the phase of a project calls for different skill requirements of a project manager.

However, since the objective of a project participant is to achieve success within a given phase the next section discusses the ways the performance of a project is measured.

1.3 Understanding Project Success

Characterizing a construction project into 'successful' or 'failure' is in fact a difficult task. This will be clearer when we analyze the preceding statement in the backdrop of the following paragraphs.

There is no universal definition of success and there is no standard methodology to measure it (McCoy 1986).

A construction project involves multiple stakeholders and each one of them has a definition of success depending on their perspectives and objectives related to a variety of elements including technical, financial, education, social, and professional issues (Parfitt and Sanvido 1993; Lim and Mohamed 1999). Jha and Iyer (2004) even go to the extreme stating that success for one participant may be the failure for other. Parfitt and Sanvido (1993) opine that failures and successes are the relative terms and they are highly subjective. The definition of success or failure can even change from project to project.

Freeman and Beale (1992) explain the disparity in points of view in the following words.

> An architect may consider success in terms of aesthetic appearance, an engineer in terms of technical competence, an accountant in terms of dollars spent under budget, a human resources manager in terms of employee satisfaction and chief executive officers rate their success in stock market.

Liu and Walker (1998) also echo the similar view as given below.

> Project success is a topic that is frequently discussed and yet rarely agreed upon. The concept of project success has remained ambiguously defined. It is a concept which can mean so much to so many different people because of varying perceptions, and leads to disagreements about whether a project is successful or not.

The perception related to the success or failure of a project may also be dependent on the time. For example a project at the time of execution may be regarded as a failure because of time and cost overrun, or due to the occurrence of fatal accidents, however, the project may be regarded a successful one if it brings development of the area, better employment, prosperity in terms of an increase in the property value of the residents, and better living conditions. Similarly a project perceived to be successful at the time of execution may be seen as a failure at the time of production/occupation.

In spite of the difficulties involved in defining the success of a project, some researchers have tried to define the success of a project. Parfitt and Sanvido (1993) quote the definition of overall success of project given by de Wit, which is given below:

> The project is considered an overall success if the project meets the technical performance specifications and/or mission to be performed, and if there is a high level of satisfaction concerning the project outcome among: key people in the parent organization, key people in the project team, and key users or clientele of the project effort.

Traditionally, success is defined as the degree to which project goals and expectations are met and the project requirements are fulfilled. However, modern projects involving multiple designers, contractors, subcontractors, construction managers, consultants, and specialists from different disciplines, and increasing domain of project requirements have compounded the problem further and understanding the success of project has become all the more complicated.

1.4 Criteria for Project Performance Evaluation

In the previous section, the difficulties in defining the success and failure of a project were discussed. It is also true that without a measurement and evaluation tool, improvement in performance cannot be made. Thus it is essential to have some criteria or performance indicators through which performance of a project can be measured and if needed performance of two projects can be compared. The measurement of project performance is often with respect to some criteria or key performance indicators (KPIs). Lim and Mohamed (1999) define criteria as the set of principles or standards by which judgement is made and are considered to be the rule of the game.

One of the most widely used project performance measure has been the 'iron triangle' consisting of schedule, cost, and quality at the three vertices (Atkinson 1999). With passage of time, other criteria have been also proposed to measure a project's performance. These performance measures can be characterized into objective and subjective categories. In the objective criteria we have tangible and measurable performance measures such as: *schedule; cost; quality; safety;* and *dispute* while in the subjective criteria we have *client satisfaction; contractor satisfaction;* and *project management team satisfaction*. A summary of the success criteria used by different researchers is presented in Table 1.1.

Freeman and Beale (1992) advocate the employment of objective measurement system to show how successfully the cost targets of the project have been achieved. They further argue that a discounted cash flow (DCF) approach is superior to other popular financial measures, partly because of its objectivity. De Wit (1988) remarks on the ability to objectively measure project success, when he states:

Table 1.1 Summary of success criteria

Author	Success criteria
Maloney (1990)	Time and cost, quality and productivity/efficiency
Norris (1990)	Budget/financial performance/profitability
Freeman and Beale (1992)	Technical performance, efficiency of project execution, managerial and organizational implications, personal growth, project termination, technical innovativeness, manufacturability, and business performance
Parfitt and Sanvido (1993)	Time and cost, budget/financial performance/profitability, health and safety, quality, meeting technical performance, specification and functionality, satisfaction of client/customer, contractor, project manager/team satisfaction, expectation/aspiration of client/contractor/project manager/team and satisfaction
Songer and Molenar (1997)	Budget, schedule, meets specification, conforms to user's expectation, high quality of workmanship, and minimizes construction aggravation.
Ashley et al. (1987)	Cost, schedule, quality, safety, and participant's satisfaction
Shenhar et al. (1997)	Project efficiency, impact on the customer, direct and business success, and preparing for the future
Lipovetsky et al. (1997)	Meeting design goals; benefits to the customer; benefits to the developing organization; and benefits to the defense and national infrastructure

> Measuring success is complex and a project is hardly ever a disaster or failure for all stakeholders during all phases in the project life cycle. A project can be a success for one party and disaster for another. (Also), a project may be perceived a success one day and failure the next. Therefore, to think that one can objectively measure the success of a project is an illusion.

Lim and Mohamed (1999) distinguish between the macro and micro view point of project success and have suggested that two criteria are sufficient to determine the macro viewpoint of project success: completion and satisfaction, whereas the completion criterion alone is enough to determine the micro viewpoint of project success.

Researchers have developed a number of frameworks to define the project success. Shenhar et al. (1997) propose, a multidimensional universal framework to assess project success based on data collected from 127 projects and evaluate relative importance of the four success criteria: *project efficiency*, *impact on the customer*, *direct and business success*, *preparing for the future* (refer Table 1.1) used in their study. They conclude that the relative importance of the success criteria may change with time and is contingent on the specific stakeholder.

Similar to above Lipovetsky et al. (1997) also find the relative importance of four chosen dimensions of success in a study of 110 defense projects performed by Israeli industry. The dimensions chosen by them are *meeting design goals*; *benefits to the customer*; *benefits to the developing organization*; and *benefits to the defense and national infrastructure* (Table 1.1). Based on the response of the three participant groups (the customer, the developing organization, and the coordinating office within the Ministry of Defense) on the dimensions, they find that *benefits to the customer* is the most important success dimension followed by *meeting design goals*. The other two dimensions are observed to be of less importance.

Baccarini (1999) uses the logical framework method (LFM) as a foundation for defining project success. Baccarini (1999) has identified four levels of project objectives: goal, purpose, output, and input. Songer and Molenaar (1997) and Chan et al. (2002) have probed the suitability of the success criteria framework for specific case of design/build projects and find that success criteria are more or less similar even in the case of design/build projects.

It is important to observe from Table 1.1 that time, cost, and quality are the most widely used criteria and can be taken as general in nature, while other criteria are project specific, e.g. Shenhar et al. (1997) and Lipovetsky et al. (1997) have concentrated their research on defense projects, and Songer and Molenaar (1997) have dealt with the success criteria of design/build projects.

Besides identifying the performance evaluation criteria and standards or the limits to measure project success have been defined by researchers. A summary of these criteria and standards are summarized in Table 1.2.

Project management as defined by (PMBOK 2000) is the application of knowledge, skills, tools and techniques to a broad range of activities to meet the requirements of the particular project. Munns and Bjeirmi (1996) define project management as the process of controlling the achievement of the project objectives. Utilizing the existing organizational structure and resources, it seeks to manage the project by applying a collection of tools and techniques, without

Table 1.2 Summary of suggested criteria and standards to measure project success

	Parameters to measure success	Suggested evaluation standards
1	On budget	The project is completed at or under the contracted cost (Songer and Molenaar 1997). The cost success criterion could be measured in terms of cost over/under run as a percentage of initial budgets (Might and Fischer 1985).
2	On schedule	The project is completed on or before the contractual finish time (Songer and Molenaar 1997). The time success criterion could be measured in terms of over/under runs as a percentage of the initial plan (Might and Fischer 1985)
3	High quality of workmanship	The completed project meets or exceeds the accepted standards of workmanship in all areas (Songer and Molenaar 1997). In other words, the project must produce what it said it would produce (PMI 1996). Quality typically includes such measures as the amount of rework required
4	Stake holder satisfaction	The completed project meets or exceeds the stakeholders' goals. The key stakeholders during the project management process are the client and the project team (Munns and Bjeirmi 1996).
5	Safety	The project honors health and safety rights of the people involved with the project by ensuring safe working condition. According to Crane et al. (1999) it can be measured by compiling safety statistics such as lost time incidents etc.
6	Dispute	The project is completed with least number of litigations resulting from disagreements among participants.

adversely disturbing the routine operations of the company. The function of project management includes defining the requirement of work, establishing the extent of work, monitoring the progress of the work, and adjusting deviations from the plan. Project management knowledge and practices are best described in terms of their component processes. These processes can be placed into five process groups (initiating, planning, executing, controlling, and closing) and nine knowledge areas (project integration management, project scope management, project time management, project cost management, project quality management, project human resource management, project communications management, project risk management, and project procurement management).

A few researchers try to distinguish project success criteria from project management success criteria(Cooke-Davies 2002; de Wit 1988). They argue that project success is to be measured against the overall objectives of the project which is a long-term and real objective of the project. The performance evaluation on the widespread and traditional measures: cost, time and quality are the measure of project management performance and they are actually short-term objective of the project. According to Baccarini (1999), project management success has three components: (a) meeting time, cost and quality objectives (project outputs and inputs); (b) quality of the project management process; and (c) satisfying project stake holder's needs where they relate to the project management process.

1.5 Need for the Study

Construction in India is the second largest industry next to agriculture and it provides employment to about 31.46 million personnel comprising both skilled and unskilled workers, technicians, foremen, clerical staff, and engineers (Jha 2011). The investment in infrastructure has doubled to US $ 500 million in the last 10 years and in the next five years plans are to secure investment of US $ 1 trillion. A large number of mega projects are under execution and planning stages. Although projects such as Delhi Metro and Delhi International Airport are perceived to be successes, the performances of large chunk of projects are not at the desired level. Successive reports of Ministry of Statistics and Programme Implementation (Infrastructure and project monitoring division) reveal poor performances of projects on time and cost accounts. According to the current report of the Ministry, 301 delayed projects have accounted for a cost overrun of Rs. 300.58 billion (US $ 7.5 billion at an exchange rate of Rs 40 for $1), i.e. 26.09 % with respect to their original sanctioned cost during the January–March, 2007 quarter (http://www.mospi.gov.in). Concerns are also raised frequently on the poor quality, safety, and dispute records in the Indian construction projects.

Indian construction companies are facing tough competition with multinational companies at present. This was not the case earlier. A large number of infrastructure facilities are being created and development works are undertaken to provide a comfortable atmosphere where construction industry has got a great role to play. It is therefore far more important to understand the current problems with the overall performance of the construction projects and suggests remedial measures appropriate to keep pace with the required growth.

Studies conducted in the developed countries have revealed a number of variables/attributes that affect the outcome of a project. 'Coordination among project participant' has been identified as one of the very important success attributes. The success of many large projects like the Petrochemical project of Reliance, Delhi Metro Project, and Delhi International Airport Project of India, multi- billion dollar Atlanta Metro Rail Project and World Trade Center Project, USA have been achieved due to better coordination among various participants of the project (Lammie and Shah 1980; Ruchelman 1980). However, there are instances when the project has failed due to lack of coordination among the participants of a project such as SCOPE Project, India and a host of large building projects in China where cost overrun is recorded to be over 50 % (Wang 2000).

However, the term 'success' itself has undergone sea change in the complex project environment with so many stakeholders. Success for one participant may be the failure for other participant. Denver airport project of USA reveals that what is viewed as a failure today may be treated as a success in future (Griffith et al. 1999). Besides, the construction projects today are no longer confined to a single discipline but are generally multidisciplinary. Modern projects involve multiple players such as number of designers, contractors, subcontractors, construction managers, consultants, and specialists from different disciplines. In a multi-agency environment, it is natural to have clash of objectives among different participants. The objective of project management is to ensure success of the project, which is not just managing the schedule, cost, and quality, generally known as 'the iron triangle'. Apart from 'the iron triangle' a number of performance measuring parameters/criteria are cited to call a project successful, such as satisfaction of project participants, technical performance of the project, and number of disputes at the completion of projects. Thus the measurement of performance also depends to a great extent on the criteria employed to measure it (PMBOK 2000).

Review of literature on the project performance reveal that most of the studies have been taken up in the context of developed countries and not enough research work has been reported for Indian construction projects. India is not yet a developed country and the performances of the projects have also not been encouraging. It is realized that more awareness among the construction professionals needs to be created. This awareness should be about project success or failure attributes, which could be taken care of while execution of a project to achieve the required outcome. It is considered that exploiting the success attributes to its maximum and minimizing the impact of the failure attributes would bring in the best results.

Also it is seen in the previous section the importance of coordination in a project has been emphasized by different case studies, but not much scientific studies are available and it still remains an underdeveloped area of managerial function. Perhaps the cause of its underdevelopment can be attributed to its having a low profile. Grigg (1993) says, "Whoever saw a news article about a conflict successfully worked out before it occurred? Planners and coordinators are just not regarded as heroes." Any country can ill afford the wastage of scarce resources on construction projects due to poor coordination. In India too, due to a number of upcoming developmental projects of large magnitude, multidisciplinary in nature and to bring improved efficiency among domestic contractors, a need for a detailed study on coordination is felt necessary.

The above reasons have been the motivating factor for the study and as the first step the various project success/failure attributes applicable to Indian construction projects are identified. Also identified are series of coordination activities that improve the project success and then the following objectives are set for the study.

1.6 Research Objectives

The present study has the following main objectives.

1. To identify critical factors responsible for success or failure of projects.
2. To evaluate the relative impact of the critical factors on the performance of project.
3. To identify the predictor variables used to predict the performance of the construction project based on schedule, cost, quality, and dispute and to

develop a user interactive model to predict the performance of the construction project.
4. To test the hypothesis that 'project success' is influenced by 'success traits' and to explore the impact of the success traits on project success.
5. To evaluate impact of coordination and coordination related activities on success of projects.
6. To study issues involved in coordination and to find out the required traits of a project coordinator.

Besides the above main objectives, there are a few other objectives that are stated in the appropriate chapters along with the main objectives.

1.7 Organization of the Book

The book is presented in NINE chapters. The remaining EIGHT chapters of the book are organized as follows.

In Chap. 2, the research method is discussed. The difficulty in getting the data of completed projects in India and necessity of adopting questionnaire survey approach for the study are described. The questionnaire development, and survey responses obtained in the study are discussed in this chapter. It also contains data analysis techniques used in the study. It contains details on univariate and multivariate analysis techniques. Techniques such as Factor analysis, multinomial regression, multinomial logistic regression, artificial neural network, and structural equation models used in the study are described in simple terms so that the readers can understand the subsequent chapters easily.

In Chap. 3, major attributes of project performance are discussed. The relative importance of the project success and failure attributes are evaluated both individually and in a collective form represented by their latent, intrinsic and common properties on four project evaluation criteria viz., schedule, cost, quality, and no-dispute.

In Chap. 4, evaluation of relative importance of different factors has been performed corresponding to the four performance criteria using multinomial logistic regression. Validation of the research findings through case studies of live projects and structured interviews with key professionals is also presented.

In Chap. 5, the above exercise has been repeated. An artificial neural network based project performance prediction model has been developed based on few predictor variables. An illustration of a user interactive project performance prediction model has also been provided.

In Chap. 6, a hypothesis that 'project success' is influenced by 'success traits' has been tested using a statistical tool called structural equation modeling (SEM). 'Success traits' was defined as a second-ordered construct composed of two latent variables: the human factors and management actions. Here the emphasis is on exploring the impact of the success traits on project success.

In Chaps. 7 and 8, various coordination activities for success of the project and allied issues in coordination are discussed. They include identification and evaluation of important coordination activities, elements affecting coordination efforts, and traits of a good project coordinator.

Finally, the present research work and the conclusions drawn in various chapters are summarized in Chap. 9. At the end the limitations of the present study are highlighted and the suggestions for further studies are given in this chapter.

References

Ashley, D., Jaselskis, E., & Lurie, C. B. (1987). The determinants of construction project success. *Project Management Journal, 18*(2), 69–79.

Atkinson, R. (1999). Project management: cost, time and quality, two best guesses and a phenomenon, its time to accept other success criteria. *International Journal of Project Management, 17*(6), 337–342.

Baccarini, D. (1999). The logical frame work method for defining project success. *Project Management Journal, 30*(4), 25–32.

Bonnal, P., Gourc, D., & Lacoste, G. (2002). The life cycle of technical projects. *Project Management Journal, 33*(1), 12–19.

Chan, A. P. C., Scott, D., & Lam, E. W. M. (2002). Framework of success criteria for design/build projects. *Journal of Management in Engineering, ASCE, 18*(3), 120–128.

Cooke-Davies, T. (2002). The 'real' success factors on projects. *International Journal of Project Management, 20*(3), 185–190.

Crane, T. G., Felder, J. P., Thompson, P. J., Thompson, M. G., & Sanders, S. R. (1999). Partnering measures. *Journal of Management in Engineering, ASCE, 15*(2), 37–42.

de Wit, A. (1988). Measurement of project success. *International Journal of Project Management, 6*(3), 164–170.

Freeman, M., & Beale, P. (1992). Measuring project success. *Project Management Journal, 23*(1), 8–17.

Griffith, A. F., Gibson, G. E., Hamilton, M. R., Tortora, A. L., & Wilson, C. T. (1999). Project success index for capital facility construction projects. *Journal of Performance of Constructed Facilities, ASCE, 13*(1), 39–45.

Grigg, N. S. (1993). New paradigm for coordination in water industry. *Journal of Water Resources Planning and Management, ASCE, 119*(5), 572–587.

Jha, K. N., & Iyer, K. C. (2004). Critical factors affecting schedule performance of Indian construction projects. *Proceedings of 4th International Conference on Construction Project Management (ICCPM 2004)* (pp. 71–80). Singapore: Nanyang Technological University.

Jha, K. N. (2011). *Construction project management: theory and practice*. New Delhi: Pearson Education.

Lammie, J. L. Shah, D. P. (1980). Project management—pulling it all together. *Journal of Transportation Engineering Division, ASCE, 116* (TE4), 437–451.

Lim, C. S., & Mohamed, M. Z. (1999). Criteria for project success: an exploratory reexamination. *International Journal of Project Management, 17*(4), 243–248.

Lipovetsky, S., Tishler, A., Dvir, D., & Shenhar, A. (1997). The relative importance of project dimensions. *R&D Management, 27*(2), 97–106.

Liu, A. M. M., & Walker, A. (1998). Evaluation of project outcomes. *Construction Management and Economics, 16*(2), 209–219.

Maloney, W. F. (1990). Framework for analysis of performance. *Journal of Construction Engineering and Management, ASCE, 116*(3), 399–415.

McCoy, F. A. (1986). Measuring success: establishing and maintaining a baseline (pp. 47–52). Montreal: Project Management Institute Seminar/Symposium.
Might, R. J., Fischer, W. A. (1985). The role of structural factors in determining project management success. *IEEE Transactions on Engineering Management, EM-32*(2), 71–77.
Munns, A. K., & Bjeirmi, B. F. (1996). The role of project management in achieving project success. *International Journal of Project Management, 14*(2), 81–87.
Norris, W. E. (1990). Margin of profit: teamwork. *Journal of Management in Engineering, ASCE, 6*(1), 20–28.
Parfitt, M. K., & Sanvido, V. E. (1993). Checklist of critical success factors for building projects. *Journal of Management in Engineering, ASCE, 9*(3), 243–249.
Pinto, J. K., Slevin, D. P. (1988a). Project success: definitions and measurement techniques. *Project Management Journal, 19*(1), 67–72.
Pinto, J. K., Slevin, D. P. (1988b). Critical success factors across the project life cycle. *Project Management Journal, 19*(3), 67–74.
Pinto, J. K., & Slevin, D. P. (1989). Critical success factors in R&D projects. *Research Technology Management, 32*(1), 31–35.
PMBOK (2000). A guide to the project management body of knowledge. Project Management Institute, http://www.pmi.org.
Project Management Institute PMI. (1996). *A guide to the project management body of knowledge*. Pa: Upper Darby.
Radosavljevic, M., & Bennett, J. (2012). *Construction management strategies: a theory of construction management*. London: Wiley-Blackwell.
Ruchelman, L. (1980). Coordinating tall building development. *Journal of the Urban Planning and Development Division ASCE, 106*(UP1), 89–101.
Shenhar, A. J., Levy, O., & Dvir, D. (1997). Mapping the dimensions of project success. *Project Management Journal, 28*(2), 5–13.
Songer, A. D., & Molenaar, K. R. (1997). Project characteristics for successful public-sector design-build. *Journal of Construction Engineering and Management, ASCE, 123*(1), 34–40.
Spitz, C. J. (1982). *The project leader: a study of task requirements, management skills and personal style*. USA: Doctoral Dissertation Case Western Reserve University.
Wang, Y. (2000). Coordination issues in Chinese large building projects. *Journal of Management in Engineering, ASCE, 16*(6), 54–61.

Chapter 2
Research Method

Abstract Research method to address the objectives (mentioned in the previous chapter) of the study is described briefly. Absence of structured data on Indian construction projects has led us to adopt questionnaire survey. Questions asked in the survey are described and the details of respondents and the participating organizations are provided. The analysis tool primarily consisted of univariate and multivariate analysis. Univariate analysis consisted of finding mean and conducting t test. In the multivariate analysis, factor analysis, multinomial logistic regression, and structural equation modeling has been utilized. These tools are discussed in sufficient details for ease in understanding the subsequent chapters. For project performance prediction model, artificial neural network has been used which is also discussed in this chapter.

2.1 Introduction

It was pointed out in the previous chapter that a number of works from the developed countries such as the USA, UK and Singapore has been carried out for the identification of success and failure factors of projects, although both success and failure aspects are not discussed together in any study. The role of project coordination in achieving success has also been concluded by different researchers. It is observed that researchers have concluded based on the case studies in their respective countries or on their personal experiences and the conclusions are specific in nature related to the case study or experience. It is obvious that these results cannot be generalized. In the Indian and other developing countries' context no such similar research has been reported.

This chapter deals with the methodology to achieve the research objectives laid down in Chap. 1. The research method broadly involves three steps: (1) Questionnaire survey approach, (2) Data analysis of responses using univariate and multivariate analyses techniques, and (3) Validation of results through case study. The schematic representation of the research method is given in Fig. 2.1.

K. N. Jha, *Determinants of Construction Project Success in India*,
Topics in Safety, Risk, Reliability, and Quality 23, DOI: 10.1007/978-94-007-6256-5_2,

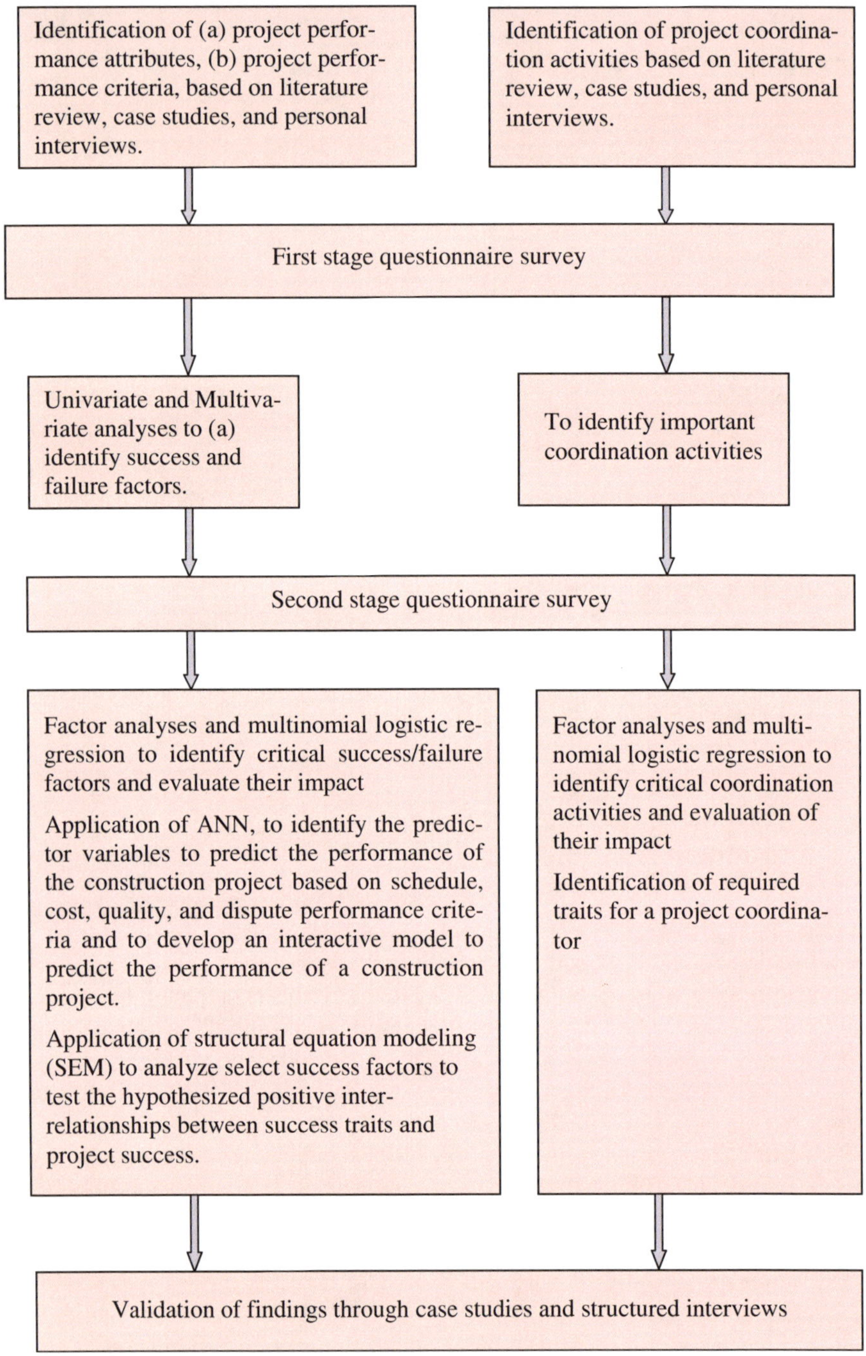

Fig. 2.1 Schematic representation of research method

2.2 Need for Questionnaire Survey

The objectives set for this research and described in the previous chapter required a large set of data on project performance to derive any meaningful conclusion. Although there are agencies which collect data on central sector projects there is no central repository which collects project performance data for both public and private projects in India. When construction companies were approached individually the issue of confidentiality was brought to the notice of the author. In cases where the companies were willing to share data, it was found that they were not in the structured form as needed for this research. It was also learnt in the process that companies would be willing to provide general information on projects.

The other route of collecting data was to conduct structured interviews of experienced professionals, however, finding an appropriate time for conducting the interview is always problematic besides being time consuming more so in the Indian context being such a vast country. There is always a chance that interviewer's bias may influence the data collection exercise.

It was for the above cited reasons that conducting a questionnaire survey to collect the data was found most apt for this research. A number of researchers have adopted this approach in the past successfully in the construction management field. Another advantage with this method is that it can cover large geographical area and thus ensuring true representative analysis.

In the present study the questionnaire survey approach was adopted in two stages: (a) the first stage, as data exploratory for detailed study, and (b) the second stage, for data collection relating to specific research issues as will be discussed later in the chapter.

2.3 Development of First Stage Questionnaire

The first stage questionnaire contains 13 broad questions seeking variety of information as given below along with the reasons for including such question in the questionnaire.

2.3.1 *Respondent's Personal Details and Professional Experience*

Personal details (see Box 2.1) and professional experiences of the respondents are asked in Q1–Q6 (see Box 2.2). These questions are asked to ensure that respondents with adequate experience and expertise only respond. It was pre-decided that if any respondent had very little experience, his response might not be considered for analysis. Further respondents' personal details might help in segregating the response sets in various categories, such as owner and contractor representatives, if required during analysis.

Box 2.1 Personal Details Questions

Name

____

Title ______________________________

Organization

____

Box 2.2 Professional Experience Related Questions

1. Length of your experience in the industry________Years

2. From among various areas of works given alongside, please cross the area of work (one only) in which you have the longest experience?

Roads		Large industrial projects	
Piling job		Tunneling Job	
Bridges		Multi-storeyed buildings	
Jetties		Factories	

3. Please indicate from the business sectors given alongside in which your business prefers to operate the most (Please cross one only).

General Building		Hydrocarbon	
Transportation(Roads)		Power, telecommunication	
Industrial Process		Manufacturing	
Ports/Harbors		Environmental	

4. Your involvement as a coordinating person was more in (Please cross one only)

Domestic operations		Overseas operation	

5. What size (in terms of average contract value) of the project were you involved in as coordinating person? (Please cross one only)

Less than Rs. 5 Crores		Between Rs. 50-100 Crores	
Between Rs. 5-10 Crores		More than Rs. 100 Crores	
Between Rs. 10-50 Crores			

6. Among the options given, which is your most preferred client? (Please cross one only).

Public Bodies		Private bodies	
Public Sector		Trust	
Multi National Companies		Societies	

2.3.2 Project Details and Traits of a Project Coordinator

Details pertaining to project details and traits of a project coordinator are sought through several sub-questions in Q7. This question is mainly meant to compare and analyze the response sets of different project details and understand the level of coordination achieved as well as to evaluate the coordinator's personal trait helping in achieving the project objective. The details sought are given below.

- Name, location, cost, duration, and contract type of the case projects (see Box 2.3).

Box 2.3 Project Details

7. Kindly furnish the details of two projects of your choice, one of which in your view wassuccessful and the other a failure.

Project data	Project 1 (Successful)		Project 2 (Failure)	
Name of the Project (Optional) Location:				
Project Cost (Rs. in Crores)	Original		Original	
	Revised		Revised	
	Achieved		Achieved	
Project Duration (in months)	Original		Original	
	Revised		Revised	
	Achieved		Achieved	
Contract Type	Turnkey		Turnkey	
	Lump sum		Lump sum	
	Item Rate		Item Rate	
	Cost Plus		Cost Plus	
	BOT		BOT	

- Repondent's involvement with the project, viz., Planning stage, Execution stage, and Operation stage (see Box 2.4).

Box 2.4 Repondent's Involvement with the Project

	Project 1 (Successful)		Project 2 (Failure)	
Your involvement was mostly in	Planning stage		Planning stage	
	Execution stage		Execution stage	
	Operation stage		Operation stage	

- Respondent's role in these projects, such as an architect, a contractor, a consultant, a subcontractor, a vendor, an owner, or a project manager (see Box 2.5).

Box 2.5 Respondent's Role

	Project 1 (Success-ful)		Project 2 (Failure)	
Your involvement was mostly as	Architect		Architect	
	Contractor		Contractor	
	Consultant		Consultant	
	Subcon-tractor		Subcontractor	
	Vendor		Vendor	
	Owner		Owner	
	Project manager		Project manager	

- Details of the number of agencies the respondent's had to coordinate directly, number of personnel involved in the project from respondent's side, number of personnel involved in the project from owner side, number of persons involved in coordination from respondent's side and number of persons involved in coordination from owner side (see Box 2.6).

Box 2.6 Coordination Issues

	Project 1 (Successful)					Project 2 (Failure)				
Number of agencies involved with you in the project (please cross one only)	0-15	15-30	30-45	45-60	>60	0-15	15-30	30-45	45-60	>60
You had to coordinate directly with how many agencies	0-15	15-30	30-45	45-60	>60	0-15	15-30	30-45	45-60	>60
Number of personnel involved in the project from your side										
Number of personnel involved in the project from owner side										
Number of persons involved in coordination from your side										
Number of persons involved in coordination from owner side										
You had to directly coordinate with ___ number of persons of other agencies										

• Time and effort expended for achieving coordination (see Box 2.7).

Box 2.7 Coordination Time and Effort

	Project 1 (Successful)		Project 2 (Failure)	
Time and effort expended for achieving coordination-i.e. the frequency of meetings held for coordination purpose-(please cross in the appropriate box)	Daily Meetings		Daily Meetings	
	Weekly Meeting		Weekly Meeting	
	Fortnightly Meetings		Fortnightly Meetings	
	Monthly Meetings		Monthly Meetings	
	Unscheduled Meetings		Unscheduled Meetings	

- Rating of the project in terms of performance on quality, dispute encountered and safety (see Box 2.8).

Box 2.8 Project Performance Rating

	Project 1 (Successful)				Project 2 (Failure)			
How do you rate the quality of the project on a scale of 1 to 4?	1	2	3	4	1	2	3	4
	1: Better than contractual requirement 2: As per contractual requirement 3: Poorer than contractual requirement 4: V Poor compared to contractual requirement							
What was the nature of dispute encountered for the project?	1	2	3		1	2	3	
	1: Major dispute 2: Minor dispute 3: No dispute							
How was the performance on safety account?	1	2	3	4	1	2	3	4
	1: Better than contractual requirement 2: As per contractual requirement 3: Poorer than contractual requirement 4: V Poor compared to contractual requirement							

- Questions on 24 traits of the person coordinating the said project (see Box 2.9).

Box 2.9 Traits of the Coordinator(Jha and Iyer 2006, reprinted with permission from Taylor and Francis)

Project data (contd..)	Project 1 (Successful)					Project 2 (Failure)				
How do you describe the traits of the person coordinating the said project	(Please <u>cross</u> the appropriate box below).					(Please <u>cross</u> the appropriate box below)				
Timeliness	1	2	3	4	5	1	2	3	4	5
Maintaining records	1	2	3	4	5	1	2	3	4	5
Interpersonal skill	1	2	3	4	5	1	2	3	4	5
Relationship with client, consultant	1	2	3	4	5	1	2	3	4	5

Box 2.9 (continued)

Coordination for achieving quality	1	2	3	4	5	1	2	3	4	5
Liaisoning skill	1	2	3	4	5	1	2	3	4	5
Knowledge of project finance	1	2	3	4	5	1	2	3	4	5
Communication skill	1	2	3	4	5	1	2	3	4	5
Reliance on systematic approach	1	2	3	4	5	1	2	3	4	5
Understanding of contract clauses	1	2	3	4	5	1	2	3	4	5
Monitoring skills	1	2	3	4	5	1	2	3	4	5
Planning skills	1	2	3	4	5	1	2	3	4	5
Forecasting skills	1	2	3	4	5	1	2	3	4	5
Facilitating skills	1	2	3	4	5	1	2	3	4	5
Resource utilization skills	1	2	3	4	5	1	2	3	4	5
Belief in team playing spirit	1	2	3	4	5	1	2	3	4	5
Analytical skills	1	2	3	4	5	1	2	3	4	5
Concern for other's ego	1	2	3	4	5	1	2	3	4	5
Concern for conciliation	1	2	3	4	5	1	2	3	4	5
Motivating skills	1	2	3	4	5	1	2	3	4	5
Follow up quality	1	2	3	4	5	1	2	3	4	5
Concern for safety, health, and welfare of labour and employees	1	2	3	4	5	1	2	3	4	5
Technical knowledge of the subject	1	2	3	4	5	1	2	3	4	5

2.3.3 *Relative Importance of Various Project Stages*

Generally a project life cycle is divided in three broad stages: planning, execution, and operation; but their relative importance within a project may vary with the type of project. The respondents are asked to rank the various stages in order of their importance and also the rank in order of importance of coordination effort required in Q8.

2.3.4 *Relative Importance of Project Performance Evaluation Parameters*

Performance of any project is generally measured by the compliance to the five parameters: budgeted schedule, budgeted cost, quality, safety, and least dispute.

Variations from the target values of above parameters by way of overruns in time and cost; non-adherence to required quality and safety norms and longstanding and increasing number of disputes are considered poor performance rating. On the other hand saving in time and cost, adherence to quality and safety standards, project participants satisfaction due to least disputes are considered successful in the performance rating. However, while evaluating the project performance, the professionals may not give equal weightage to all these parameters and these parameters are given subjective ratings due to project complexities of various nature and varied experiences of professionals. In order to get a generalized value of the weightages, Q9 is framed (see Box 2.10). It is assumed that mean value of large responses to this question would give the relative preference in a generalized form for these parameters.

Box 2.10 Preference of Success Criteria

Past studies have indicated that the following five main criteria are used to measure the project success. Their relative importance/weightage to assess the overall project success varies from individual to individual and project to project. Please rank them in the decreasing order of their importance for evaluating project success in column (3), for example rank 1 means the most important criteria followed by 2 for the next important and so on. Similarly to achieve the requirement as given in column (2), the extent of coordination efforts would also vary. Please rank the extent of coordination efforts required for each criterion requirement in column (4), for example rank 1 means the highest amount of coordination effort required followed by 2 for the next higher amount of coordination effort and so on.

(1)	(2)	(3)	(4)
S.No.	Success Criteria	Ranking for evaluating project success(choose from 1 to 5)	Rank of extent of coordination effort required (choose from 1 to 5)
1	Compliance to 'sched-ule'		
2	Compliance to 'cost'		
3	Compliance to 'quality'		
4	Compliance to 'least dispute'		
5	Compliance to 'safety'		

2.3.5 *Project Performance Attributes*

Based primarily on literature review and partly on the basis of personal interviews of experienced construction professionals a list of 55 project attributes affecting the project outcome in terms of 'success' and 'failure' were compiled. These were listed in Q10(a) of the questionnaire. A sample question showing the manner in which the responses on these attributes were planned to be captured is shown in Box 2.11 and the description of 55 attributes are given in Table 2.1. The list of 55 attributes was first published in Iyer and Jha (2005). Although the list cannot be termed as exhaustive, yet it covered current understanding of most of the attributes known to affect project outcome. In order to capture any significant attribute which the author might have missed, an open ended question [Q10(b)] was included in the questionnaire. Schedule, cost, quality, and dispute were the four performance measuring criteria used in the questionnaire. The five-point scale used for capturing the respondents' opinion is included at the bottom of Box 2.11.

Table 2.1 Project success attributes (Iyer and Jha 2005, with permission from Elsevier)

No.	Project success attributes
1	Size and value of the project being large
2	Scope and nature of work well defined in the tender
3	Aggressive competition at tender stage
4	Urgency emphasized by the owner while issuing tender
5	Inadequate project formulation in the beginning
6	Uniqueness of the project activities requiring high technical know-how
7	Favorable political and economic environment
8	Hostile political and economic environment
9	Hostile social environment
10	Favorable social environment
11	Harsh climatic condition at the site
12	Favorable climatic condition at the site
13	Project completion date specified but not yet planned by the owner
14	Monitoring and feedback by client
15	Timely decision by the owner or his engineer (reluctance or otherwise)
16	Understanding operational difficulties by the owner engineer thereby taking appropriate decisions
17	Top management's enthusiastic support to the project manager (PM) and project team at site
18	Top management's backing up the plans and identify critical activities
19	Selection of PM with proven track record at an early stage by top management
20	Delegating authority to project manager by top management
21	Monitoring and feedback by top management
22	Reluctance in timely decision by top management
23	Effective monitoring and feedback by PM
24	Effective monitoring and feedback by the project team members
25	Lack of understanding of operating procedure by the PM
26	Ignorance of appropriate planning tools and techniques by PM
27	Reluctance in timely decision by PM
28	Ability to delegate authority to various members of his team by PM

(continued)

Table 2.1 (continued)

No.	Project success attributes
29	Coordinating ability and rapport of PM with his team members and sub-contractor
30	Coordinating ability and rapport of PM with top management
31	Coordinating ability and rapport of PM with owner representatives
32	Coordinating ability and rapport of PM with other contractors at site
33	Leadership quality of PM
34	Project manager's authority to take financial decision, selecting key team members, etc.
35	Project manager's technical capability
36	Construction control meetings
37	Regular budget update
38	Commitment of all parties to the project
39	Understanding of responsibilities by various project participants
40	Authority to take day to day decisions by the PM's team at site
41	Conflicts among team members
42	Conflicts between PM and top management
43	Conflicts between PM and other outside agency such as owner, sub-contractor or other contractors
44	Tendency to pass on the blame to others
45	Availability of resources (funds, machinery, material, etc.) as planned throughout the project duration
46	Developing and maintaining a short and informal line of communication among project team
47	Poor human resource management and labour strike
48	Presence of crisis management skill of PM
49	Vested interest of client representative in not getting the project completed in time
50	Training the human resources in the skill demanded by the project
51	Mismatch in capabilities of client and architect
52	The capability of project participants to market the end product to the intended users
53	Positive attitude of PM, and project participants
54	Negative attitude of PM, and project participants
55	Holding key decisions in abeyance

2.3.6 Coordination Activities

In the previous chapter it has been observed that researchers have identified 'coordination among project participants' as one of the important variables responsible for the success of many projects and lack of coordination among the participants of a project has been responsible for failure of projects. Taking lead from the number of coordination activities identified by Saram and Ahmed (2001) a preliminary list of coordination activities that best suited Indian construction was prepared. The preliminary list was later modified through personal interviews with professionals and a pilot survey was then conducted. The response to the pilot survey helped in identifying a number of new coordination activities and improving the presentation of questionnaire. In total 59 coordination activities are identified. Q11 seeks opinion on contribution of these 59 coordination activities in achieving the project objectives. While a sample question is shown in Box 2.12, the description of 59 coordination activities is provided in Table 2.2.

Box 2.11 Project attributes

(a) Listed below are some of the attributes responsible for advantages/hindrances to project success. Please indicate the effects of these attributes on various project success evaluation criteria given alongside the attributes.

S.No.	Project success attributes	Effect on completion schedule					Effect on project cost					Effect on project quality					Effect on Project Dispute				
1	Size and value of the project being large	1	2	3	4	5	1	2	3	4	5	1	2	3	4	5	1	2	3	4	5
2	Scope and nature of work well defined in the tender	1	2	3	4	5	1	2	3	4	5	1	2	3	4	5	1	2	3	4	5
	... See Table ___for the complete list of success attributes...																				

Legend:

1	Adversely delay	Adversely affect	Adversely affect	Adversely increase
2	Significantly delay	Significantly affect	Significantly affect	Significantly increase
3	Marginally delay	Marginally affect	Marginally affect	Marginally increase
4	No effect	No effect	No effect	No effect
5	Helps speedingup progress	Helps in saving	Helps in improving	Helps in decreasing

Importance of these coordination activities is also gaged on a five-point scale separately in four project performance criteria: schedule, cost, quality, and no-dispute. Here '1' refers to 'Very Large', '2' refers 'Large', '3' refers 'Small', '4' refers 'Very small', and '5' refers 'Unnoticeable' effect on the mentioned success criteria.

Box 2.12 Importance of coordination activities (Jha and Iyer 2006, with permission from Elsevier)

Listed below are the coordination activities that one needs to carry out for better outcome of project. Please indicate the effects of these activities on various project success evaluation criteria given alongside these activities.
NOTE:-Any activity you feel to be "not applicable" with reference to coordination you may score them by putting a line across those activities and noting 'NA'; and while some other important activities you feel to be missing may be added at the end of Table.

S.No.	Description of coordination activities	Extent of positive effect on																			
		completion schedule					project cost					project quality					non occurrence of project Dispute				
1	Implementing all contractual commitments	1	2	3	4	5	1	2	3	4	5	1	2	3	4	5	1	2	3	4	5
2	Arranging for timely carrying out of all tests for inspections and approval by the engineerand maintaining records of the same.	1	2	3	4	5	1	2	3	4	5	1	2	3	4	5	1	2	3	4	5

Legend: 1: Very large 2: Large 3: Small 4: Very small 5: Unnoticeable

It is well established that schedule, cost, quality, safety, and no-disputes are the five important project evaluation criteria. Among these criteria, "safety" provisions are statutory requirements, and it has to be met by all means. Any project manager would automatically put in all necessary coordination efforts to achieve these requirements. However, the variation in achievement of other criteria is possible due to various reasons; hence the other four criteria are only selected for evaluation of importance of coordination parameters. The respondents are further told to strike off those activities, which they feel are not related to coordination and add the activities they feel to be missing.

2.3.7 Parameters Influencing Coordination

Q12 seeks response on criticality of factors affecting coordination and their influence on coordination effort required. For the development of Question 12 (see Box 2.13) the study conducted by Hastak and Shaked (2000) have proved to be helpful.

Table 2.2 Description of coordination activities (Jha and Iyer 2006, with permission from Elsevier)

S. No.	Description of coordination activities
1	Implementing all contractual commitments
2	Arranging for timely carrying out of all tests for inspections and approval by the engineer and maintaining records of the same
3	Arranging submission of samples of materials for approval by the engineer
4	Reporting progress reports, resources deployment report etc. as required by the engineer
5	Providing storage space, testing facilities, scaffolding, plant, power, water, illumination, etc. to other agencies as envisaged by the contract
6	Arranging for compliance with site instructions/directives from the engineer and revising programs/ordering material accordingly
7	Applying good technical practices
8	Preparing a project quality plan in line with contract specification
9	Communicating instances of poor quality, dangerous, or adverse incidents/situations to relevant personnel
10	Caring for works of others by making staff and workmen aware of their responsibilities in this regard
11	Coordinating hand over of work areas/service areas (such as plant rooms, service routes, etc.) to other parties
12	Proposing remedial work methods and programs for executing in case of defect or damage
13	Identification of appropriate human resources, materials, and equipment for the project
14	Estimating the optimum resource requirements
15	Proper assignment of task to the available human resources for the project
16	Organizing resources (manpower, plant, and material) for effective utilization.
17	Arranging technical and behavioral training of human resources
19	Facilitating payments to own employees and subcontractors
18	Managing the health, safety, and welfare of employees
20	Managing the maintenance and safety of plant and machinery
21	Equipping own men and subcontractors with tools, equipment, and resources
22	Explaining and supporting the work of nominated subcontractors and specialist suppliers
23	Delegation of responsibilities to appropriate project participant
24	Regular follow up of work delegated to project participant
25	Ensuring discipline among all employees
26	Resolving differences/conflicts/confusion among participants
27	Motivating project participants
28	Developing a team spirit and receiving constructive input from all participants in the project
29	Identifying/gathering information on requirements of all parties and consolidating for use in planning
30	Identification of activities on critical path
31	Communicating project progress, financial and commercial status, plans, schedules, changes, documents, etc., to all relevant participants
32	Regular monitoring of critical path activities for adhering to schedule
33	Coordinating the purchases, delivery, storage and handling of materials

(continued)

Table 2.2 (continued)

S. No.	Description of coordination activities
34	Arranging for kick off meeting and review with all departments asking for date wise schedule for their area of activities
35	Identifying or gathering information on defects, deficiencies, ambiguities, and conflicts in drawings and specifications and having them resolved
36	Improving/altering/eliminating activities and considering better alternatives that may efficiently meet the project objectives
37	Arranging inputs like drawings, specifications, and technical details on time for execution
38	Providing an organized means for gathering information and compiling records
39	Identifying and gathering information on project work requirements (grouting, openings, making good, etc.) of all relevant parties and coordinating the time and manner of their execution
40	Preparing coordination drawings for freezing sequence of activities and giving a road map of responsibilities to all involved in the project
41	Agreeing on detailed methods of construction with all the parties involved
42	Coordinating and rescheduling the sequence of onsite work in case of changes in requirement from client side
43	Interfacing/integrating the work on different subsystems
44	Establishing and maintaining an effective organizational structure and communication channels
45	Conducting regular meetings and project reviews
46	Analyzing the project performances on time, cost and quality, detecting variances from the schedule/requirements and dealing with their effects considering time and resource constraints
47	Monitoring the budget on all activities and taking corrective action
48	Monitoring the overall functioning of each section and department of the project
49	Keeping joint records of all drawings, amendments to contract, directives, correspondences, verbal instructions, and documents received from the project participants (consultants, clients and vendors etc.)
50	Keeping joint records of quantities of work done especially of the work that is to get covered up
51	Keeping joint records of price escalations where the contract has escalation clause
52	Keeping joint records of owner supplied materials along with their scheduled delivery dates and actual receipt date
53	Keeping joint records of all input cost (viz. labour, material, plant etc.) for non tendered items
54	Keeping joint records of adverse weather conditions, breakdown time of client supplied equipment etc.
55	Coordinating with offsite fabricators and their deliveries
56	Acting as liaison with specialist consultants, specialist subcontractors, nominated subcontractors etc.
57	Maintaining proper relationships with client, consultants, and the subcontractor
58	Acting as liaison with the client and the consultants
59	Contacting outside authorities for testing, inspection and approval etc.

2.4 Development of Second Stage Questionnaire

The second stage questionnaire is developed based on the analysis results of the first stage questionnaire to get specific information on certain issues. For example, the analyses of responses on the project success and failure attributes in the first stage questionnaire have identified the attributes in 20 distinct groups that demonstrate certain latent properties of the attributes called "factors".

Through the second stage questionnaire, attempt is made to measure the importance of these factors at different levels of project performances. Similarly, the top 20 coordination activities identified from the responses of the first stage

Box 2.13 Factors affecting coordination

12. What according to you are the factors/parameters in order of their criticality (on a scale of 1 to 5 based on the given legend) and their influence on coordination effort required (again on a scale of 1 to 5 as per the legend).

S.No.	Parameter Description	Parameter characteristics					Influence on coordination effort				
1	Increasing project size (in terms of contract price)	1	2	3	4	5	1	2	3	4	5
2	Increasing project duration	1	2	3	4	5	1	2	3	4	5
3	Nature of project from regular job to a more complex job	1	2	3	4	5	1	2	3	4	5
4	Increasing project complexity (e.g. complex architectural features)	1	2	3	4	5	1	2	3	4	5
5	Very high degree of hazard associated with the project	1	2	3	4	5	1	2	3	4	5
6	Lack of PM's (Project Manager) experience in handling the project/work	1	2	3	4	5	1	2	3	4	5
7	Contract type - from conventional one (such as item rate, cost plus, lump sum) to complex ones (turnkey, BOT etc.)	1	2	3	4	5	1	2	3	4	5
8	Inadequate period of completion given in the contract	1	2	3	4	5	1	2	3	4	5
9	Inadequate drawings and details	1	2	3	4	5	1	2	3	4	5
10	Presence of liquidated damage clause in contract	1	2	3	4	5	1	2	3	4	5
11	Absence of liquidated damage clause in contract	1	2	3	4	5	1	2	3	4	5
12	Previous experience with the associated agencies viz., contractors/ clients/consultants	1	2	3	4	5	1	2	3	4	5
13	The type of client - from Government bodies to Private bodies	1	2	3	4	5	1	2	3	4	5
14	Presence of labour union	1	2	3	4	5	1	2	3	4	5
15	Increasing restriction in working hour	1	2	3	4	5	1	2	3	4	5
16	The location of the project from the native states to other states/countries.	1	2	3	4	5	1	2	3	4	5
17	Selection of knowledgeable and motivated subcontractors	1	2	3	4	5	1	2	3	4	5
18	Presence of reliable subcontractor	1	2	3	4	5	1	2	3	4	5
19	Requirement of stage passing i.e. examination and approval of owner before commencing next process of work in a multi stage work	1	2	3	4	5	1	2	3	4	5
20	Non-availability of finance for the project	1	2	3	4	5	1	2	3	4	5
21	Supply of resources like material and equipment in the client' scope	1	2	3	4	5	1	2	3	4	5
22	Increasing complexity of business sector from conventional (e.g. buildings, roads, bridges) to more advanced business sector (e.g. tunnelling and jetties etc.)	1	2	3	4	5	1	2	3	4	5
23	Delay due to involvement of many agencies/statuary bodies for approval	1	2	3	4	5	1	2	3	4	5

Legend :

Parameter characteristics		Influence on coordination effort	
1	Highly critical	1	Significant increase
2	Critical	2	Marginal increase
3	Important but not critical	3	Marginal decrease
4	Not so important	4	Significant decrease
5	Irrelevant	5	Inconsequential

questionnaire are further scrutinized through a different set of questions. In this questionnaire the respondents are asked to select a project of their own choice (referred to as 'choice project' in the study) with which they have been fully associated or they have complete knowledge about the project. Respondents are asked to base their responses on the choice project. The second stage questionnaire contains eight questions and seeks information on the following.

2.4.1 Personal and Project Details

Through Q1–Q5 (see Box 2.14) the personnel details of the respondents like their names, affiliation, experience, etc. are asked first and details of the choice project, such as the name, location, cost, schedule, and the type of the choice project are also requested.

Box 2.14 Personal and Project Details

Name ______________________

Designation ______________________

Organization ______________________

Length of your experience in the industry ________Years

Please answer the following questions keeping in mind ANY ONE CONSTRUCTION PROJECT (Case Project) of your choice. (The project could be anything: successful or failure.)

1. Name of the Case Project (Optional)______________________
2. Location: ______________________
3. Project Cost (Rs. in Crores) ______________________
4. Project Duration (in months) ______________________
5. Type of project (e.g. Roads, Industrial projects, Buildings, Heavy engineering, etc.)

Q6 seeks respondents' view on the overall performance of the choice project as well as performance in terms of adherence to schedule, cost, quality, no-dispute and safety (see Box 2.15). Responses on the performance were sought on 10-point scale where 1 represents poor performance with minimum score and 10 represents very good performance with maximum score.

Box 2.15 Performance of Case Project

6. How do you rate the performance of the case project in terms of overall performance and separately performance on schedule, cost, quality, dispute and safety account on a scale of 1 to 10 given in the legend below.

1	2	3	4	5	6	7	8	9	10
Poor			Average		Fair		Good		Very Good

S.No.	Performance Measure	Your Evaluation									
I	Overall performance of the project	1	2	3	4	5	6	7	8	9	10
II	Performance on schedule account	1	2	3	4	5	6	7	8	9	10
III	Performance on cost account	1	2	3	4	5	6	7	8	9	10
IV	Performance on quality account	1	2	3	4	5	6	7	8	9	10
V	Performance on dispute account	1	2	3	4	5	6	7	8	9	10
VI	Performance on safety account	1	2	3	4	5	6	7	8	9	10

2.4.2 Project Performance Factors

As discussed above, the 20 success and failure factors (identified from the analysis of first stage questionnaire responses) are listed out in Q7 (see Box 2.16). Respondents are asked to rate contribution of these factors on the actual outcome of the choice project in all the four performance levels. Since the outcome of the factor could have positive effect toward enhancing the performance or negative causing decline, the responses are sought on a 11-point scale ranging from −5 to +5 (−5 indicating most negative effect, +5 indicating most positive effect and 0 meaning no effect).

Box 2.16 Project Performance Factors

7. There are many variables that affect the outcome of any project. Through a previous study we have been able to identify the following factors, which are a combination of different variables and has been defined in the appendix. How do you rate the extent to which these factors contributed in the outcome of the Case project?

--------------------Extent of contribution--------------------										
-5	-4	-3	-2	-1	0	1	2	3	4	5
High				Low	No effect	Low				High

S.No.	Factors affecting the outcome of the project	Your View (Please cross in the appropriate box)										
1	Project manager's competence	-5	-4	-3	-2	-1	0	1	2	3	4	5
2	Top management support	-5	-4	-3	-2	-1	0	1	2	3	4	5
3	Monitoring and feedback by project participants	-5	-4	-3	-2	-1	0	1	2	3	4	5
4	Favorable working condition	-5	-4	-3	-2	-1	0	1	2	3	4	5
5	Commitment of all project participants	-5	-4	-3	-2	-1	0	1	2	3	4	5
6	Owners competence	-5	-4	-3	-2	-1	0	1	2	3	4	5
7	Interaction between project participants-internal	-5	-4	-3	-2	-1	0	1	2	3	4	5
8	Interaction between project participants-external	-5	-4	-3	-2	-1	0	1	2	3	4	5
9	Good coordination between project participants	-5	-4	-3	-2	-1	0	1	2	3	4	5
10	Availability of trained resources	-5	-4	-3	-2	-1	0	1	2	3	4	5
11	Regular budget update	-5	-4	-3	-2	-1	0	1	2	3	4	5
12	Conflict among project participant	-5	-4	-3	-2	-1	0	1	2	3	4	5
13	PM's ignorance and lack of knowledge	-5	-4	-3	-2	-1	0	1	2	3	4	5
14	Hostile socio economic environment	-5	-4	-3	-2	-1	0	1	2	3	4	5
15	Owner's incompetence	-5	-4	-3	-2	-1	0	1	2	3	4	5
16	Indecisiveness of project participants	-5	-4	-3	-2	-1	0	1	2	3	4	5
17	Harsh climatic condition at site	-5	-4	-3	-2	-1	0	1	2	3	4	5
18	Aggressive competition during tendering	-5	-4	-3	-2	-1	0	1	2	3	4	5
19	Negative attitude of project participants	-5	-4	-3	-2	-1	0	1	2	3	4	5
20	Faulty project conceptualization	-5	-4	-3	-2	-1	0	1	2	3	4	5

2.4.3 Important Coordination Activities

The last question Q8 seeks information on the contribution of the 20 important coordination activities in achieving proper coordination (see Box 2.17). The respondents are asked to rate the extent to which these activities were carried out in the choice project on a scale of 1–5 with 1 indicating "unsatisfactorily done" to 5 indicating "excellently done".

Box 2.17 Important Coordination Activities (Jha and Iyer 2006, with permission from Elsevier)

8. Through a previous study we have identified the following 20 important coordination activities that play significantly in the outcome of any project. To what extent were these coordination activities actually performed in the case project? Please select from 1 to 5 as given in the legend.

Legend:

1	Unsatisfactorily done
2	Fairly done
3	Satisfactorily done
4	Nicely done
5	Excellently done

1	Implementation of all contractual commitments	1	2	3	4	5
2	Arrangement for timely carrying out of all tests for inspections and approval by the engineer and maintaining records of the same.	1	2	3	4	5
3	Arranging submission of samples of materials for approval by the engineer	1	2	3	4	5
4	Application of good technical practices	1	2	3	4	5
5	Preparation of a project quality plan in line with contract specification	1	2	3	4	5
6	Arranging remedial work methods and programs for executing in case of defect or damage	1	2	3	4	5
7	Identification of appropriate human resources, materials and equipments for the project	1	2	3	4	5
8	Estimation of the optimum resource requirements	1	2	3	4	5
9	Proper assignment of task to the available human resources for the project	1	2	3	4	5
10	Organization of resources (manpower, plant, and material) for effective utilization.	1	2	3	4	5
11	Ensuring discipline among all employees	1	2	3	4	5
12	Resolving differences /conflicts /confusion among participants	1	2	3	4	5
13	Motivation of project participants	1	2	3	4	5
14	Development of a team spirit and receiving constructive input from all participants in the project	1	2	3	4	5
15	Identification of activities on critical path	1	2	3	4	5
16	Regular monitoring of critical path activities for adhering to schedule	1	2	3	4	5
17	Arrangement of required inputs like drawings, specifications, and technical details on time for execution	1	2	3	4	5
18	Agreement on detailed methods of construction with all the parties involved	1	2	3	4	5
19	Analysis of the project performances on time, cost and quality, detecting variances from the schedule/requirements and dealing with their effects considering time and resource constraints	1	2	3	4	5
20	Monitoring the overall functioning of each section and department of the project	1	2	3	4	5

2.5 Responses Received

2.5.1 First Stage Questionnaire

The organized segment of Indian construction industry comprises of about 250 large (more than 500 employees) and 500 medium (between 200 and 500 employees) organizations besides 25,000–30,000 small (less than 200 employees) organizations. The unorganized industry segment (stand-alone contractors) comprises 120,000 firms (CFI 2005).

The target population consisted of 750 large and medium organizations and out of this a total of 400 respondents were identified from the addresses available with government offices, Builders Association of India and through personal contacts. The first stage questionnaires were mailed to them in the last week of August 2002. The professionals included owner representatives (both active in service and retired) and contractors from 51 top and medium organizations. The names of the responding organizations are given in Table 2.3. A total of 114 responses were

Table 2.3 Names of the responding organizations for first stage questionnaire

Contractors	Owners	Consultants
ACC Ltd	Border Road Organization	Ashok Arora Associates
AFCONS	Bihar State Electricity Board	Batra & Associates
Ahluwalia Contracts Private Ltd	Construction Industry Development Council	Bhardwaj & Bhardwaj Ltd
Bagai Construction Private Ltd	Central Public Works Dept.	Building Material Testing Promotion Council
Bhageeratha Engineering Ltd	Delhi Metro Rail Corporation	Central Water Commission
Continental Construction Ltd	Chief Architects Organization J&K Govt	Consulting Engineering Services
Gammon India Ltd	Army Corps of Engineers	Dalal Mott Macdonald Private Ltd
Gannon Dunkerley & Co. Ltd	GraminVikas Special Division	Engineers India Limited
International Metro Civil Contractors	National Thermal Power Corporation	Hospital Services Consultancy Corporation
L&T ECC Division	Himachal Pradesh State Electricity Board	Institute of Construction Project Management
Mitsui Kensetsu India Ltd	Indian Railway Construction Corporation	RITES Ltd
New Delhi Construction Private Ltd	Mathura Brindawan Development Authority	Spazio Design Architecture Private Ltd
N S Construction	Military Engineers Services	Vintech Consultants
Punj Lloyd Ltd	New Delhi Municipal Corporation	WAPCOS Chief Engineer
Siemens	National Highway Authority of India	
Skanska	Reserve Bank of India	
Wig Brothers Construction Private Ltd	National Hydroelectric Project Corporation	
Builders Association of India	UP State Bridge Corporation	
	Water Resources Department	

received which included 42 owner representatives (the government officials), 28 consultants, and 41 contractors. The names of the organization in three responses were missing. The response rate was 28.5 %.

2.5.2 Second Stage Questionnaire

From the target population consisting of 750 large and medium organizations, the sample size was kept at 40 % (300) for the second stage questionnaire, anticipating a response rate of 25–30 %, which is considered normal in a questionnaire survey. For this case also, the respondents were selected randomly from across the country, identified from the addresses available with Government offices, Builders Association of India, and through personal contacts. The respondents had wide range of experience both in number of years of service and in handling variety of projects. Only those respondents who had served a minimum of 10 years in the construction industry and had executed a contract value of at least Rs. 50 million were selected. No threshold value of turnover of respondent's organization was, however, specified to ensure wider participation. A total of 91 responses were received out of the 300 mailed. The response received was around 30 % of the sample and can be considered reasonable, given the length of the questionnaire and target respondents (Krosnick 1999). In total 93 % of respondents had worked in domestic projects while 7 % had worked in overseas projects. The contract value of the projects reported by the respondents varied between Rs. 50 million and 1,000 million (1 USD = Rs. 45) and a mixture of Turnkey, Lump-sum, and Item rate contracts in general building construction, transportation projects, power projects, bridges, factories, and refinery projects.

A summary of respondents' profile is given in Table 2.4 from where it can be seen that respondents having an experience between 10–20 years form the largest group and the average years of experience of the respondents work out to be 18 years. These respondents also have varied experience of handling contracts of different sizes. Also shown in the table is the summary of respondents' areas of expertise both in terms of work area as well as business area.

For second stage questionnaire 300 questionnaires were sent to randomly selected professionals from top contracting organizations of the country. The Questionnaires were sent to them in April 2003. A total of 92 responses were received which included 30 owner representatives (the government officials), 62 contractors. The response rate was 30.7 %. Names of the responding organizations are given in Table 2.5.

The analysis of the second-stage questionnaire survey identified the impact of the 20 factors mentioned earlier. In the survey, the respondents had to choose a project, which they had executed. This was referred to as the 'choice project' in the study. For the 'choice project', the respondents rated the extent of contribution of the 20 factors on an 11-point scale (−5 to +5, with −5 indicating high negative contribution, 0 being no effect, and +5 indicating high positive contribution) on the

Table 2.4 Summary of respondent's profile

Experience in years	Percentage (%)	Contract size	Percentage (%)	Work area	Percentage (%)	Business sector	Percentage (%)
Less than 10	18	Less than INR 50 million+	31	Roads	30	General building	53
Between 10–20	40	Between INR 50–100 million	19	Large industrial projects	21	Petroleum	10
Between 20–30	21	Between INR 100–500 million	22	Piling jobs	3	Transportation	16
More than 30	21	Between INR 500–1000 million	8	Tunneling jobs	6	Power	7
		More than INR 1000 million	20	Bridges	8	Industrial process	5
				Multistoried buildings	27	Manufacturing	2
				Factories	5	Telecom	1
						Environmental	5

Table 2.5 Names of the responding organizations for second stage questionnaire

Contractors	Owners	Consultants
BRPL	Army Corps of Engineers	Central Water Commission
Continental Construction Ltd	Border Road Organization	Consulting Engineering Services
Dual Structural and Industries Ltd	BHEL	CSW
GEA Energy System	BSES Rajdhani Power Ltd	Engineers India Limited
GEA Cooling Tower Tech. (I) Pvt. Ltd	Delhi Development Authority	RITES Ltd
KSHI JV	Development Authority Kanpur	TECHNIP Ltd
L&T ECC Division	Indian Railway Construction Corporation	
Modular Pvt. Ltd	IFFCO	
NMRD Ltd	Mazgon Dock Ltd	
OMAXE Construction Ltd	Military Engineers Services	
Punj Lloyd Ltd	ONGC	
Siemens	SAIL	
TEXCEL Engineers Private Ltd		
Top Line Build Tech Private Ltd		

schedule performance of the project. The respondents also rated the schedule performance of the 'choice project' on a 10-point scale (1 being very poor performance and 10 being very good performance).The respondents included the owners, the contractors, and the consultants from the government, the public sector, and multinational companies conducting business in India.

2.6 Analysis Techniques

Responses for the questionnaire are given identification number for easy identification and retrieval. All the responses are stored and analyzed using SPSS (Rel 9). The analyses method used in the study included univariate and multivariate analysis, artificial neural network, and structural equation modeling. Brief description and utility of various analyses tools are given below.

2.6.1 Univariate Analyses

The details of various univariate analyses used in the study are briefly described below:

1. *Summary statistics* of responses—consisted of mean, standard deviation, and frequency.

2. *t* test—The t-tests are used for two purposes, namely to compare the sample means with the corresponding hypothesized mean values, and to test the significance of differences between the means of two independent samples.
3. *Analysis of variance (ANOVA)*—This was used mainly to evaluate the difference in opinion of the owners and contractors on success attributes and coordination activities. Chapters 3 and 4 contain extensive uses of this technique.
4. *Non parametric Spearman's rank correlation technique*—mainly to measure the association between two ranked variables where data are not available to use in numeric form, but information is sufficient to rank the data first, second, third and so forth. These ranks are then used to calculate rank correlation coefficients between pairs of ranks, which indicate the measure of correlations (degrees of association) that exists between pairs. The analyses results using this technique are provided in Chap. 7.

2.6.2 Multivariate Analyses

Multivariate analyses in the present study comprise of two techniques: factor analysis and multinomial logistic regression. The combination of these two methods has been used successfully in the study to identify the critical success and failure factors and the critical coordination activities. The details of the two methods and their applications are discussed briefly below:

1. *Factor analysis*—This has been mainly used to identify success and failure factors out of 55 project attributes. The applications are illustrated in Chaps. 4 and 5. Factor analysis provides a method for combining variables based on their concomitant variation and uncovers a common underlying property among a set of correlated variables, called a 'factor'. This technique has been employed successfully in different disciplines including construction management (Maloney and McFillen 1995). According to Hair et al. (2006), factor analysis is used primarily to identify set of dimensions that are latent (not easily observed) in a large set of variables. In other words it is a method of combining or condensing large number of people into distinctly different groups within a larger population. The factors so obtained can partially or completely replace the original set of variables and can be used in subsequent regression, correlation or discriminant analysis. Some of the commonly used terms in factor analysis are: factor loadings, common factor vectors, rotation of factors, communality, and factor scores. These are explained beautifully in Overall and Klett (1972), Dillon and Goldstein (1984), Child (1990), and Hair et al. (2006).
2. *Multinomial Logistic Regression*—This method of regression is used when the dependent variables are with more classes. It is also used when the independent variables are either continuous or categorical or both. The questionnaire design and the scale selection in this study was tailor made for the application of multinomial logistic regression. Whitehead (1998) lists out three problems in

using ordinary least square in such circumstances. They are the problem of non normal distribution of error, likelihood of predicted probabilities becoming greater than 1 or less than 0, and the heteroskedastic error terms. Some commonly used terms in multinomial logistic regression are: logit coefficients, Wald statistics, and maximum likelihood estimation. These are described in detail in Whitehead (1998), and reference manuals of NCSS and SPSS software. The multinomial logistic regression model is simply a non linear transformation of the linear regression and it calculates changes in the log odds of the dependent.

2.6.3 Artificial Neural Network

The ANN has been used successfully in many fields including construction management. ANN model has been used in this study to identify the predictor variables to predict the performance of the construction project based on schedule, cost, quality, and dispute performance criteria and to develop an interactive model to predict the performance of a construction project.

An ANN model is a mathematical model in which the information processing occurs in a number of simple elements called neurons (nodes); signals are transmitted between neurons over connection links that have an associated weight with them. Each neuron applies an activation function to the incoming signal to determine its output signal (Zurada 1992).

Figure 2.2 shows a binary threshold unit proposed by McCulloch and Pitts. This mathematical model computes a weighted sum of its '*n*' input signals $X_j = 1, 2, \ldots$, '*n*' and generates an output of '1' if it is above a certain threshold or else gives an output of '0' (cited in Zurada 1992). The most commonly used activation function is the sigmoid, the other being linear and Gaussian (Jain et al. 1996).

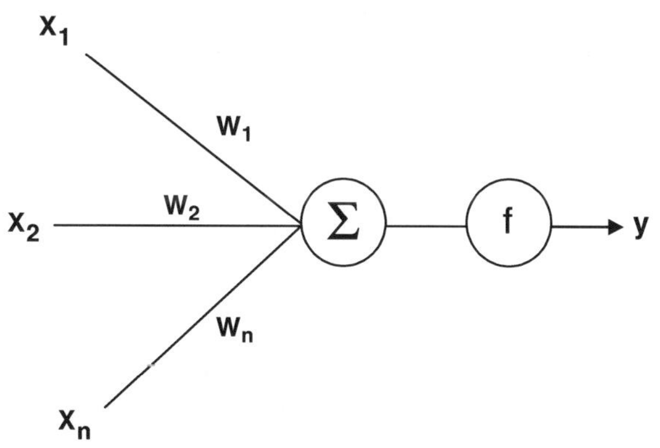

Fig. 2.2 McCulloch—Pitts model of a neuron (Jha and Chockalingam 2011, with permission from Taylor and Francis)

To express mathematically, each *neuron j* sums its weighted input as follows:

$$netj = \sum_{j=1}^{n} w_j \times x_j \quad (2.1)$$

The output of a *neuron*, *y* is a function of its weighted input, expressed as follows:

$$y = f(netj) \quad (2.2)$$

In its most simple form, an ANN model has only two layers- an input layer and an output layer, although an ANN model can also be constructed with more than one hidden layer and each hidden layer may also have different numbers of neurons. The input layer is connected to the input parameters through a weighted factor. The weights represent the strength of the connections to a unit which can be altered through learning rules to create a desired input/output response from an artificial neural network. Different network architectures require appropriate learning algorithms and the ANN's ability to automatically learn through training process makes them attractive and user friendly. There are several learning algorithms and Jain et al. (1996) report that the back propagation learning algorithm with feed forward network architecture is most suited for predictions.

ANN is classified into feed forward network and recurrent network based on the connection pattern. In the former there are no loops while in the latter the loops occur because of feedback connections.

2.6.4 Structural Equation Modeling

The structural equation modeling (SEM) has been used in this study to analyze select success factors to test the hypothesized positive inter-relationships between success traits and project success. The SEM is a statistical technique and can be regarded as the extension of multivariate techniques, most notably multiple regression and factor analysis. It combines a measurement model (confirmatory path analysis) and a structural model (regression or path analysis) in a single statistical test (Kline 1998; Mueller 1996; Garver and Mentzer 1999). Molenaar et al. (2000) opine that many of the research issues in construction engineering and management can be dealt with effectively using the SEM. The SEM comprises of (a) a measurement model concerned with how well the variables measure the latent factors addressing their reliability and validity and (b) a structural model concerned with modeling the relationships between the latent factors by describing the amount of explained and unexplained variance (Wong and Cheung 2005).

The steps involved include reliability assessment, exploratory factor analysis, confirmatory factor analysis, and validity tests. The first step in SEM is the validation of measurement model through confirmatory factor analysis (CFA). The CFA allows for the assessment of fit between observed data and a priori conceptualized, theoretically grounded model that specifies the hypothesized causal relationships between constructs and their observed indicator variables (Mueller and Hancock 2004). The assessment of the overall model fit is carried out using multiple goodness-of-fit (GOF) indices, including the ratio of Chi square to degree of freedom, the root mean square error of approximation (RMSEA), the comparative fit index (CFI), and the non normed fit index (NNFI). The assessment of overall model fit is considered a critical issue in the SEM. In the study, LISERAL 8.8 has been used as structural equation modeling (SEM) tool.

References

CFI. (2005). *Gearing construction for growth. Construction Federation of India*, Jan 2005.

Child, D. (1990). *The Essentials of Factor Analysis*. London: Cassell Educational Limited.

Dillon, W. R., & Goldstein, M. (1984). *Multivariate analysis*. New York: Wiley.

Garver, M. S., & Mentzer, J. T. (1999). Logistics research methods: Employing structural equation modeling to test for construct validity. *Journal of Business Logistics, 20*(1), 33–57.

Hair, J. F., Black, W. C., Babin, B. J., Anderson, R. E., & Tatham, R. L. (2006). *Multivariate data analysis* (6th ed.). Upper Saddle River: Pearson Prentice Hall.

Hastak, M., & Shaked, A. (2000). ICRAM-1: Model for international construction risk assessment. *Journal of Management in Engineering, ASCE, 16*(1), 59–69.

Iyer, K. C., & Jha, K. N. (2005). Factors affecting cost performance: evidence from Indian construction projects. *International Journal of Project Management, 23*(4), 283–295.

Jain, A. K., Mao, J., & Mohiuddin, K. M. (1996). Artificial neural networks: A tutorial. *IEEE (theme feature), 24*(3), 31–44.

Jha, K. N., & Iyer, K. C. (2006). Critical determinants of project coordination, *International Journal of Project Management*, *24*(4), 314–322

Jha, K. N., & Chockalingam, C. T. (2011). Prediction of schedule performance of Indian construction projects using an artificial neural network, *Construction Management & Economics*, Taylor & Francis, *29*(9),901–911

Kline, R. B. (1998). *Principles and practice of structural equation modeling*. New York: Guilford Press.

Krosnick, J. A. (1999). Survey research. *Annual Review of Psychology, 50*, 537–567.

Maloney, W. F., & McFillen, J. M. (1995). Job characteristics: Union-nonunion differences. *Journal of Construction Engineering and Management, ASCE, 121*(1), 43–54.

Molenaar, K., Washington, S., & Diekmann, J. (2000). Structural equation model of construction contract dispute potential. *Journal of Construction Engineering and Management, 126*(4), 268–277.

Mueller, R. O. (1996). *Basic principles of structural equation modeling*. New York: Springer.

Mueller, R. O., & Hancock, G. R. (2004). Factor analysis and latent structure: Confirmatory factor analysis. In N. J. Smelser & P. B. Baltes (Eds.), *International encyclopedia of the social and behavioral sciences* (pp. 5239–5244). Oxford, England: Pergamon.

Overall, J. E., & Klett, C. J. (1972). *Applied multivariate analysis*. New York: McGraw Hill Book Company, Inc.

Saram, D. D. D., & Ahmed, S. M. (2001). Construction coordination activities: What is important and what consumes time? *Journal of Management in Engineering, ASCE, 17*(4), 202–213.

Whitehead, J. (1998). Available at www.personal.ecu.edu/whiteheadj.

Wong, P. S. P., & Cheung, S. O. (2005). Structural equation model of trust and partnering success. *Journal of Management in Engineering, 21*(2), 70–80.

Zurada, J. M. (1992). *Introduction to artificial neural systems*. Los Angeles: West Publishing Company.

Chapter 3
Major Attributes of Project Performance

Abstract Literature review in the area of project success and failure attributes has been carried out. Project success and failure attributes identified through literature review, case studies, and personal interviews are analyzed based on the response of questionnaire survey. Project success and failure attributes are arranged in the order of their relative importance for the four performance criteria: schedule, cost, quality, and no-dispute and the top attributes in each of these criterions are discussed. Factor analyses were conducted on the project success and failure attributes separately for all the four performance criteria. This has resulted in a set of success and failure factors for all the four performance criteria. A common set of 11 project success factors and nine failure factors emerging from all the four performance criteria has also been compiled for further analysis.

3.1 Introduction

In the previous chapter research method has been discussed. In this chapter analysis of responses on influence of project success/failure attributes on project outcome are discussed. The objectives being discussed in this chapter are as given below.

- To identify the relative importance of success and failure attributes in Indian construction industry as perceived by two important stakeholders—owners and contractors in a construction contract on four performance evaluation criteria: schedule; cost; quality; and no-dispute.
- To understand the latent properties of the success and failure attributes

Project attributes are the variables responsible to influence the outcome of a project. The attributes can be people (project participants and their qualities and traits), resources, technology, working environment and system, or tasks. The genesis of many research works in this area is on the assumption that it is possible

K. N. Jha, *Determinants of Construction Project Success in India*,
Topics in Safety, Risk, Reliability, and Quality 23, DOI: 10.1007/978-94-007-6256-5_3,

to find certain success attributes and the project success is repeatable (Ashley et al. 1987). Also there are certain attributes called as failure attributes, which when present lead to failure of the project. Finding the success attributes and maximizing them is as important as finding the failure attributes and minimizing them. Accordingly researchers have put their energy in identifying the success attributes and failure attributes, with the common objective of enhancing chances of project success. These success and failure attributes are discussed separately in the next two sections.

3.2 Success Attributes

Project success/failure attributes are summarized in Table 3.1 first published in Iyer and Jha (2005). These attributes are compiled based on the existing literature pertaining to construction, research and development, defense, and development projects. Most of the studies reported in Table 3.1 focus on specific success measuring parameters and the critical success factors (CSF) identified by the researchers are either industry or contract specific.

Rockart (1982) defined critical success factors as those few key areas of activity in which favorable results are absolutely necessary for a particular manager to reach his or her own goals…those limited number of areas where things must go right.

Rowlinson (1999) on the other hand states that critical success factors are those fundamental issues inherent in the project, which must be maintained in order for team working to take place in an efficient and effective manner. According to Rowlinson (1999) the critical success factors require day-to-day attention throughout the life of the project.

Based on the analyses of a questionnaire survey, Pinto and Slevin (1988b) observe significant relation between project success and the factors believed to be critical to project success. The factors appearing in their list are: (1) project mission, (2) top management support, (3) project schedule-plans, (4) client consultation, (5) personnel, (6) technical tasks, (7) client acceptance, (8) monitoring and feedback, (9) communication, and (10) trouble-shooting.

Using the above success factors Pinto and Slevin (1989) have further developed a 10-factor model of the project implementation process and called it Project Implementation Profile. This model could be used by the project managers to periodically monitor the state of each of the factors throughout a project's life. Further they also find that the relative importance of the critical success factors change with the project life cycle stage. They also identify four additional external factors: (1) characteristics of the project team leader, (2) power and politics within the organization, (3) environmental events, and (4) the urgency of the project.

Ashley et al. (1987) conclude that construction project success is repeatable and requires a great deal of work to understand it for achieving cost effectiveness and a competitive position. They report that there exists statistically significant

Table 3.1 Summary of select studies in the area of project success/failure factors (Reprinted from Iyer and Jha 2005 with permission from Elsevier)

Author	Summary of the past works
Rubin and Seeling (1967)	Authors assess the impact of project manager's experience on the project's success/failure. They use technical performance of the project as a measure of success and conclude the following
	Project manager's previous experience has minimal impact on the project's performance and
	Size of the previously managed project does affect the manager's performance
Avots (1969)	This is a theoretical study to understand the reasons for project failure. The author concludes that choice of wrong project manager, the unplanned project termination, and unsupportive top management are the main reasons of failure
Sayles and Chandler (1971)	Authors conclude the following factors as critical for success of a project
	Project manager's competence, Scheduling, Control systems and responsibilities, Monitoring and feedback, and Continuing involvement in the project
Martin (1976)	Author concludes the following factors as critical for success of a project
	Define goals, Select project organizational philosophy, General management support, Organize and delegate authority, Select project team, Allocate sufficient resources, Provide for control and information mechanics, Require planning, and review
Baker et al. (1983)	The authors suggest that instead of time, cost and performance as the project success criteria, perceived performance should be used as the success criteria. They observe the following success factors
	Clear goals, Goal commitment of project team, On site project manager, Adequate funding to completion, Adequate project team capability, Accurate initial cost estimates, Minimum start-up difficulties, Planning and control techniques, Task (vs. social orientation), and Absence of bureaucracy
Cleland and King (1983)	Authors identify the following success factors
	Project summary, Operational concept, Top management support, Financial support, Logistic requirements, Facility support, Market intelligence (who is the client), Project schedule, Executive development and training, Manpower and organization, Acquisition, Information and communication channels, and Project review
Locke (1984)	Author identifies the following success factors
	Make project commitments known, Project authority from the top, Appoint competent project manager, Set up control mechanisms (schedules, etc.), and Progress meetings
Hughes (1986)	The author identifies that the projects fail because of improper basic managerial principles, such as the improper focus of the management system, by rewarding the wrong actions, and the lack of communication of goals
Morris and Hough (1987)	Authors identify the following success factors through a study of eight large and complex projects having great potential economic impact but poorly managed and generally failed
	Project objectives, Technical uncertainty innovation, Politics, Community involvement, Schedule duration urgency, Financial contract legal problems, and Implementation problems

(continued)

Table 3.1 (continued)

Author	Summary of the past works
Schultz et al. (1987)	Authors classify critical success factors in two groups as given below and conclude that these groups affect project performances at different phases of implementation The strategic group consisting of factors like project mission, Top management support and Project scheduling Tactical group consisting of factors like client consultation and personnel selection and training
Pinto and Slevin (1989)	Continuing the previous work of Schultz et al. (1987), these authors evaluate the relative importance of tactical group and strategic group of factors over the project life cycle. They conclude that when external success measures are employed, planning factors dominate tactical factors throughout the project life cycle
Chua et al. (1997)	Budget performance is given the primary importance in the study. Through an application of neural network approach authors identify the eight important project management attributes associated with achieving successful budget performance: (1) number of organizational levels between the project manager and craft workers, (2) amount of detailed design completed at the start of construction, (3) number of control meetings during the construction phase, (4) number of budget updates, (5) implementation of a constructability program, (6) team turnover, (7) amount of money expended on controlling the project and (8) the project manager's technical experience They also claim that their model can be used as a predictive tool to forecast budget performance of a construction project
Chan et al. (2001b)	They identify a set of project success factors for design and build (D&B) projects and examine the relative importance of these factors on project outcome. Using Factor analysis from the response of 53 participants on 31 variables they extracted six project success factors. These are project team commitment, contractor's competencies, risk and liability assessment, client's competencies, end-users' needs, and constraints imposed by end-users. Further Project team commitment, client's competencies, and contractor's competencies were found to be important to bring successful project outcome from the multiple regression findings
Schaufelberger (2003)	Well defined project scope, mutual understanding of the scope of work between the owner and the contractor, owner had sufficient experience with the design build project delivery method
Nguyen et al. (2004)	Competent project manager, adequate funding throughout the project, multidisciplinary/competent project team, commitment to project, and availability of resources
Andersen et al. (2006)	Strong project commitment, early stakeholder influence, stakeholder endorsement of project plans and rich project communications, a well-structured and formal project approach and well understood and accepted project purpose
Jha and Iyer (2007)	Top management support, owner's competence, commitment of the project participants; conflict among project participants, coordination among project participants; favorable working condition; project manager's competence; and interaction between project participants—internal and interaction between project participants—external
Lam et al. (2008)	Project nature, effective project management action, and adoption of innovative management approaches

difference between the extent of key attributes of average construction projects and that of outstanding construction projects. These key attributes are: planning effort (construction and design); project team motivation; project manager goal commitment; project manager technical capabilities; control systems; and scope and work definition.

Chua et al. (1999) through an application of neural network approach has established the association of eight important project management attributes (Table 3.1) with achieving successful budget performance. Chan et al. (2001b) analyze the six project success factors (Table 3.1) for Design and Build (D&B) Projects and identify *project team commitment*, *client's competencies*, and *contractor's competencies* to be important to bring successful project outcome. Chan et al. (2001a) advise that human aspects also need attention in addition to functional aspects of project performance. They find that inter-organizational teamwork leads to (1) successful project performance, (2) development of positive view of the D/B procurement method and (3) higher job satisfaction.

Study by Mansfield and Odeh (1991) suggest better management of human resources on construction projects by making use of motivation. Lack of motivation is one of the major reasons for the low productivity in construction industry when compared with other industries. This is despite advancement in equipment, materials, and methods of design and construction. Mansfield and Odeh further point out the complexities involved in achieving motivation of project staff due to the peculiar nature of construction industry like short term employment, working in adverse environment, facing unusual problems, nature of construction contracts and risk allocation between contractor and the owner, the fluctuation in labor supply, and the management issues on multi-contractor projects etc.

Mcneil and Hartley (1986) assert that the successful project depends on experienced personnel, good communications, and dedicated planning. They further emphasize group participation in developing the project plan; defining planning detail; determining project milestones; preparing detailed plans; and using available tools as key to achieving success. According to Thompson (1991), client's role during project development and implementation is crucial to the success of the project. Project managers, however, often they lack support from the client organization's top management.

Based on a survey questionnaire to collect data from practitioners, Nguyen et al. (2004) spelt out the success factors for large construction projects in Vietnam. They utilized factor analysis to categorize these success factors under four categories: comfort, competence, commitment, and communication and referred them as the four COMs.

Schaufelberger (2003) selected seven contractors from the Engineering News Record list of "Top 100 Design-Build-Firms" and interviewed their local project managers using a common set of questions. The study finds that contractors were most interested that the project scope was well defined, that there was a mutual understanding of the scope of work between the owner and the contractor, and that the owner had sufficient experience with the design-build project delivery method. Further the study finds that the primary determinants regarding the decision whether or not to pursue a project are: the owner's reputation regarding treatment of contractors; the size, scope, and location of the project; and the contractor's current backlog. Interestingly, honorariums do not figure in this list of primary determinants and it seemed to have little influence on some contractors' decision whether or not to pursue a project.

Andersen et al. (2006) studied the relationship between project success factors and actual project success and investigated those factors within the direct influence of project managers that can make a real difference to the outcome of project endeavors. The study applied principal components analysis (PCA) on ten project success items and three distinctively different project success criteria were extracted. The most important factors in improving managerial ability to deliver results in time and at cost were strong project commitment, early stakeholder influence, stakeholder endorsement of project plans, and rich project communications

Jha and Iyer (2007) identified 55 project performance attributes are identified and conducted a two-stage questionnaire survey. While 11 success and nine failure factors are identified from the analysis of the first stage questionnaire responses, the second stage questionnaire survey has helped in evaluation of the extent of criticality of these factors with respect to a given performance rating of the project. It is found that extent of contribution of various success or failure factors varies with current level performance ratings of the project. The crux of the findings of this study has been the emergence of commitment, coordination, and competence as the key factors for achievement of schedule, cost, and quality objectives respectively.

3.3 Failure Attributes

Avots (1969) concludes that *choice of wrong project manager*; *the unplanned project termination*; and *unsupportive top management* are the main reasons for project failure. Hughes (1986) in another study identifies that projects fail because of improper basic managerial principles such as improper focus of the management system, rewarding the wrong actions, and the lack of communication of goals. Chitkara (1998) points out inadequate project formulation and the improper management of the projects as the primary reasons for project failures.

While discussing the straight forward project life-cycle approach, Bonnal et al. (2002) say that this approach has two dominant phases: pre-project phase and project phase. While pre-project phase is characterized by creativity, project phase is characterized by rigor. These two phases require different skill set to manage them. The authors conclude that some project failures can be explained because the dichotomy of skills has not been respected.

Some of the factors causing delay and cost overrun identified by Mansfield et al. (1994) in Nigerian construction projects are: *poor contract management*, (which is caused by lack of adequate experience and training at the senior management level, and inadequate technical manpower, a very low level of productivity, inadequate finances for short-and long- term purposes and an absence of specialization), *financing and payment of completed works*, *changes in site conditions*, *shortages of materials and plant items*, *design changes*, *and subcontractors and nominated suppliers*. Contractors, consultants, and Public clients are in general agreement on the causes of delay and cost overruns. Some other factors responsible for cost overrun are price fluctuations, inaccurate estimates prepared by the contractors, delays and additional works, fraudulent practices, and kickbacks. As a remedial measures, these authors recommend giving more emphasis on thorough project analysis before authorization, ensuring the availability of adequate funds before projects are started, adoption of alternative contract methods like BOOT etc., adoption of an efficient material management system that includes harnessing local construction materials and development of manpower in the areas of project management, information and database management systems. Chan and Kumaraswamy (1997) determine and evaluate the relative importance of the significant factors causing delays in Hong Kong construction projects. They find 'poor site management and supervision', 'unforeseen ground conditions', 'low speed of decision making involving all project teams', 'client-initiated variations', and 'necessary variations of works' as the five principal and common causes of delays. Poor scope definition has been described as one of the leading causes of project failure in the U.S. construction industry (Dumont et al. 1997).

3.4 Identification of Relative Importance of Project Performance Attributes

Relative importance of project attributes are identified through the responses of Q10a of the first stage questionnaire separately for all performance parameters. The mean values of responses to all attributes are calculated and the attributes are arranged in the descending order of mean value. The scale pattern across all the performance parameters of the question generally suggests response of 5 for 'positive effect'; 4 for 'no effect'; 3 for 'marginal negative effect'; 2 for 'significant negative effect'; and 1 for 'adverse effect'. However, the mean of responses would not give a whole number as asked in the question and hence for interpretation purposes any effect is considered to lie between the mid points of two adjacent scales. Depending upon the mean scores of responses, the attributes are then segregated in three groups: the first group (with $\mu \geq 4.5$) that shows positive contribution; the second group (with $4.5 < \mu < 3.5$) which is neutral and passive having no significant impact on the project outcome and the third group (with $\mu \leq 3.5$) indicating negative impact. Accordingly attributes having $\mu \geq 4.5$ are called success attributes; attributes with μ between 4.5 and 3.5 are called neutral and attributes with $\mu \leq 3.5$ are called as failure attributes. The interpretations of these three groups in the context of different performance parameters are given in Table 3.2.

Table 3.2 Interpretations of various ranges of mean values of responses pertaining to success attributes

Type of effect	$\mu \geq 4.5$	$4.5 < \mu < 3.5$	$\mu \leq 3.5$
When schedule is the success criteria	Helps speeding up the progress	Neither positively contributing nor adversely affecting	Adversely affecting
When cost is the success criteria	Helps in saving	Neither positively contributing nor adversely affecting	Adversely affecting
When quality is the success criteria	Helps in improving	Neither positively contributing nor adversely affecting	Adversely affecting
When dispute is the success criteria	Helps in decreasing	Neither positively contributing nor adversely affecting	Adversely affecting

Depending on the performance measuring criteria the analysis has resulted in different sets of success attributes ($\mu \geq 4.5$). For example when schedule is the performance measuring criteria, a total of 31 attributes have emerged in this group and when cost is the performance measuring criteria, a total of 30 attributes have emerged in this group. Similarly for quality performance a total of 29 and for no-dispute performance criteria a total of 28 success attributes have emerged. Mean values of all the success attributes and their rank orders under the four performance criteria are summarized criterion-wise in Table 3.3.

The rank orders of the success attributes in different evaluation criteria suggest that 'Positive attitude of PM, and project participants' has emerged to be the most important attribute when schedule, quality, and no-dispute criteria are of prime importance in gauging the project performance, while 'Coordinating ability and rapport of PM with top management' takes supreme importance when cost criterion is considered.

Some of the top ranking success attributes across the four performance criteria are observed to be: Positive attitude of PM, and project participants; Proper rapport and coordination among PM, top management, owner's representatives, team members and sub-contractor; Selection of PM with proven track record at an early stage by top management; Effective monitoring and feedback by the project team members and PM himself; and leadership quality and technical ability of the PM. The top five success attributes in the four project performance criteria are reproduced in Table 3.4 for ready reference.

The attributes of group 2 ($3.5 < \mu < 4.5$) neither causing any positive contribution nor negative impact and they being neutral are dropped from the study. The attributes falling under group 3 (for $\mu \leq 3.5$) in the four performance criteria are given in Table 3.5.

It is seen that depending on the performance measuring criteria the analysis has resulted in different rank order of failure attributes. Also, when schedule is the performance measuring criteria, a total of 22 attributes have emerged in this group and when cost is the performance measuring criteria, a total of 23 attributes have

Table 3.3 Rank of success attributes ($\mu \geq 4.5$) based on performance criterion performance criterion

S. No.	Project success attributes	Mean of schedule	Rank of schedule	Mean of cost	Rank of cost	Mean of quality	Rank of quality	Mean of no-dispute	Rank of no-dispute
1	Positive attitude of PM, and project participant	4.84	1	4.74	4	4.75	1	4.81	1
2	Selection of PM with proven track record at an early stage by top management	4.82	2	4.63	8	4.56	14	4.68	7
3	Effective monitoring and feedback by the project team members	4.82	3	4.78	2	4.73	5	4.67	8
4	Authority to take day to day decisions by the PM's team at site	4.81	4	4.61	10	4.69	6	4.59	17
5	Coordinating ability and rapport of PM with his team members and sub-contractor	4.80	5	4.59	15	4.54	18	4.68	6
6	Project manager's authority to take financial decision, selecting key team members, etc.	4.75	6	4.54	16	4.58	12	4.60	16
7	Leadership quality of PM	4.75	7	4.59	13	4.74	4	4.63	14
8	Understanding of responsibilities by various project participants	4.74	8	4.59	12	4.64	8	4.65	10
9	Coordinating ability and rapport of PM with other contractors at site	4.74	9	4.52	18	4.51	19	4.68	5
10	Top management's backing up the plans and identify critical activities	4.73	10	4.50	20	4.44	23	4.56	20
11	Commitment of all parties to the project	4.73	11	4.54	17	4.59	11	4.64	11
12	Top management's enthusiastic support to the project manager (PM) and project team at site	4.72	12	4.59	14	4.60	9	4.62	15
13	Effective monitoring and feedback by PM	4.70	13	4.75	3	4.75	2	4.70	3
14	Coordinating ability and rapport of PM with top management	4.70	14	4.78	1	4.69	7	4.75	2
15	Understanding operational difficulties by the owner engineer thereby taking appropriate decisions	4.70	15	4.68	6	4.50	20	4.59	18

(continued)

Table 3.3 (continued)

S. No.	Project success attributes	Mean of schedule	Rank of schedule	Mean of cost	Rank of cost	Mean of quality	Rank of quality	Mean of no-dispute	Rank of no-dispute
16	Project manager's technical capability	4.69	16	4.70	5	4.74	3	4.58	19
17	Coordinating ability and rapport of PM with owner representatives	4.69	17	4.51	19	4.47	22	4.69	4
18	Monitoring and feedback by top management	4.68	18	4.61	9	4.55	16	4.52	23
19	Monitoring and feedback by client	4.63	19	4.33	29	4.40	24	4.34	28
20	Construction control meetings	4.62	20	4.42	23	4.58	13	4.67	9
21	Training the human resources in the skill demanded by the project	4.61	21	4.38	26	4.60	10	4.52	22
22	Delegating authority to project manager by top management	4.57	22	4.49	22	4.47	21	4.63	13
23	Favorable climatic condition at the site	4.56	23	4.35	27	4.39	27	4.27	N/A
24	Timely decision by the owner or his engineer (reluctance or otherwise)	4.54	24	4.65	7	4.54	17	4.63	12
25	Availability of resources (funds, machinery, material) as planned throughout the project duration	4.50	25	4.35	28	4.55	15	4.54	21
26	Regular budget update	4.49	26	4.50	20	4.30	28	4.51	24
27	Favorable social environment	4.46	27	4.12	N/A	4.06	N/A	4.21	N/A
28	Scope and nature of work well defined in the tender	4.46	28	4.60	11	4.40	25	4.40	26
29	Favorable political and economic environment	4.45	29	4.39	24	4.09	N/A	4.48	25
30	Ability to delegate authority to various members of his team by PM	4.35	30	4.38	25	4.29	29	4.26	N/A
31	Developing and maintaining a short and informal line of communication among project team	4.33	1	4.22	30	4.39	26	4.35	27

Table 3.4 Summary of top five success attributes in different criteria

Rank	Attribute description for schedule criterion	Attribute description for cost criterion	Attribute description for quality criterion	Attribute description for no-dispute criterion
1	Positive attitude of PM, and project participants	Coordinating ability and rapport of PM with top management	Positive attitude of PM, and project participants	Positive attitude of PM, and project participants
2	Selection of PM with proven track record at an early stage by top management	Effective monitoring and feedback by the project team members	Effective monitoring and feedback by PM	Coordinating ability and rapport of PM with top management
3	Effective monitoring and feedback by the project team members	Effective monitoring and feedback by PM	Project manager's technical capability	Effective monitoring and feedback by PM
4	Authority to take day to day decisions by the PM's team at site	Positive attitude of PM, and project participants	Leadership quality of PM	Coordinating ability and rapport of PM with owner representatives
5	Coordinating ability and rapport of PM with his team members and sub-contractor	Project manager's technical capability	Effective monitoring and feedback by the project team members	Coordinating ability and rapport of PM with other contractors at site

emerged in this group. Similarly for quality and no-dispute performances a total of 22 failure attributes have emerged. While 'Conflicts between PM and other outside agency such as owner, sub-contractor or other contractors' has the most adverse effect in achieving the schedule performance, 'Poor human resource management and labour strike' emerges to be the root cause for cost escalation at project sites. 'Negative attitude of PM, and project participants' is the prime reason for underperformance on quality account and increase in dispute. Some of the other top ranking critical attributes having adverse performance on the four performance criteria are: inadequate project formulation in the beginning; conflicts between PM and top management; mismatch in capabilities of client and architect; tendency to pass on the blame to others; holding key decisions in abeyance; conflicts among team members; and vested interest of client representative in not getting the project completed in time. The importance of these attributes revealed in this study is consistent with the findings of other researchers as can be observed from Literature review chapter. For easy reference the top five attributes under each criterion are summarized in Table 3.6.

It can be pointed here that while there are minor differences in mean values across the four performance criteria for all the success attributes (Table 3.3), the difference is considerable in case of some of the failure attributes given in Table 3.5. It is apparently supporting the view that some failure attributes have

Table 3.5 Rank of failure attributes $\mu \leq 3.5$ based on performance criterion

S. No.	Project success attributes[a]	Mean of schedule	Rank of schedule	Mean of cost	Rank of cost	Mean of Quality	Rank of quality	Mean of no-dispute	Rank of no-dispute
1	*Conflicts between PM and other outside agency such as owner, sub-contractor or other contractors*	1.54	1	1.83	6	2.54	14	1.87	2
2	*Poor human resource management and labor strike*	1.54	1	1.58	1	2.01	2	2.01	6
3	*Conflicts between PM and top management*	1.55	3	1.76	4	2.16	6	2.06	8
4	*Inadequate project formulation in the beginning*	1.56	4	1.72	2	2.35	10	1.90	3
5	Negative attitude of PM, and project participants	1.56	4	1.74	3	1.71	1	1.65	1
6	*Reluctance in timely decision by PM*	1.65	6	1.88	7	2.37	12	2.10	10
7	*Vested interest of client representative in not getting the project completed in time*	1.66	7	1.78	5	2.36	11	2.01	6
8	Conflicts among team members	1.77	8	2.00	11	2.12	5	2.12	11
9	Ignorance of appropriate planning tools and techniques by PM	1.81	9	2.00	11	2.21	8	2.23	13
10	*Hostile political and economic environment*	1.84	10	2.16	15	2.76	16	2.54	16
11	Tendency to pass on the blame to others	1.86	11	2.19	16	2.12	4	2.10	9
12	Lack of understanding of operating procedure by the PM	1.87	12	1.99	10	2.19	7	2.12	11
13	*Reluctance in timely decision by top management*	1.88	13	2.07	13	2.75	15	2.49	15
14	Mismatch in capabilities of client and architect	1.88	13	1.88	8	2.09	3	1.97	5
15	Holding key decisions in abeyance	1.91	15	1.98	9	2.30	9	1.94	4
16	*Harsh climatic condition at the site*	1.96	16	2.09	14	2.39	13	2.96	18
17	*Hostile social environment*	2.01	17	2.25	17	2.90	17	2.55	17
18	*Project completion date specified but not yet planned by the owner*	2.19	18	2.42	18	3.05	20	2.37	14
19	*Uniqueness of the project activities requiring high technical know-how*	2.75	19	2.77	19	3.29	21	3.18	21
20	Size and value of the project being large	3.32	20	3.32	21	3.58	22	3.00	20
21	*Aggressive competition at tender stage*	3.32	21	3.64	22	2.99	19	2.97	19
22	Presence of crisis management skill of PM	3.84	22	3.81	23	3.84	N/A	4.10	N/A
23	*Urgency emphasized by the owner while issuing tender*	3.85	N/A	2.91	20	2.91	18	3.31	22
24	The capability of project participants to market the end product to the intended users	4.00	N/A	4.03	N/A	4.05	N/A	4.09	N/A

[a] Attributes shown in italics have varied intensity of adverse effect on the four performance criteria

Table 3.6 Summary of top five failure attributes in different criteria

Rank	Attribute description for schedule criterion	Attribute description for cost criterion	Attribute description for quality criterion	Attribute description for no-dispute criterion
1	Conflicts between PM and other outside agency such as owner, sub-contractor, or other contractors	Poor human resource management and labour strike	Negative attitude of PM, and project participants	Negative attitude of PM, and project participants
2	Poor human resource management and labor strike	Inadequate project formulation in the beginning	Poor human resource management and labor strike	Conflicts between PM and other outside agency such as owner, sub-contractor, or other contractors
3	Conflicts between PM and top management	Negative attitude of PM, and project participants	Mismatch in capabilities of client and architect	Inadequate project formulation in the beginning
4	Inadequate project formulation in the beginning	Conflicts between PM and top management	Tendency to pass on the blame to others	Holding key decisions in abeyance
5	Negative attitude of PM, and project participants	Vested interest of client representative in not getting the project completed in time	Conflicts among team members	Mismatch in capabilities of client and architect

different level of adverse affect on all the four performance criteria. To explore this, the following hypotheses were set up.

$$\text{Null Hypothesis } H_0 : \mu_1 = \mu_2 = \mu_3 = \mu_4$$

$$\text{Alternate Hypothesis H1 } : \mu_1, \mu_2, \mu_3, \text{ and } \mu_4 \text{ are not all equal.}$$

where μ_1, μ_2, μ_3, and μ_4 are mean values corresponding to schedule, cost, quality, and no-dispute criteria.

If the attribute mean values do not differ significantly it can be inferred that the attribute in question has similar intensity of effect across all the four performance criteria. On the other hand if the differences among the mean values of attribute in question are too large to be ascribed to chance sampling error, it can be inferred that the failure attribute does have dissimilar effects on the four performance criteria. Accordingly the suitable measures can be devised to address the issue.

The hypotheses testing were conducted at 0.05 level of significance. It can be observed from Table 3.5 that there exist significant differences in the responses on

14 failure attributes (written in bold face in Col 2 of Table 3.5) among the four criteria i.e., the intensity of adverse affects in influencing the outcome in terms of four performance criteria vary for these 14 attributes. For the remaining nine attributes there is no significant difference in their mean values.

3.5 Intrinsic Characteristics of Success and Failure Attributes

While various attributes given in the questionnaire have been assumed to influence the project performances to varying degrees, it was hypothesized that response pattern would indicate that the influences could be distinctly categorized, with each influence category emerging from different sets of attributes. It was hypothesized so because various attributes listed could not be said to have unique and independent feature, but are correlated with each other and form groups. The common property among the attributes of a group, which may be latent till the group is distinctly identified, may be the influencing parameter. Thus it is realized that instead of analyzing the attributes independently, it is better to view them as groups with various attributes of the group possessing some common property, provided that such grouping is possible and reliable. As was discussed in Chap. 2, factor analysis is one such statistical technique that analyzes the data points of variables and segregates the variables under different factors—each factor possessing a group of variables with high factor loadings. The group of variables in a given factor represents some common and latent property among them that are considered to be influencing property of the set of variables represented by the factors. It is therefore customary to name the factor also after the latent and common property of the group of variables emerging in the factor. It was also pointed out in this Chapter that if the communality of each of the variables after the factor analysis is greater than or equal to 0.3 ($\geq$0.3) the factor model is considered to be reliable.

Subsequent to identification of success and failure attributes the responses received for these attributes on four performance evaluation criteria (schedule, cost, quality, and no-dispute) were segregated in two data sets called (a) success attributes data sets and (b) failure attributes data sets. These two data sets represented responses to the success attributes and responses to the failure attributes respectively. In order to know the correlations among the attributes that would indicate some latent or intrinsic properties among the correlated variables, factor analysis was employed individually on responses to success attributes and failure attributes. As it has been observed that responses are sought on four project performance criteria, there are four different response sets within the given data set and factor analysis was applied on all the four response-sets individually.

Since the responses to the questionnaire were received predominantly from two groups: owners/owner organizations and contractors/contracting organization, and their interests in any project are quite opposite (Iyer 1996), it was also hypothesized that their responses would differ significantly. Thus each response set was also segregated as owner response data set and contractor response data set and factor analysis was also applied individually for these two response sets. Since there was no significant difference between the factors emerging in these two response data sets, the combined response (also called as overall response elsewhere in the book) were only used instead of owner response data set and contractor response data set separately.

Initially, the extracted factors, which were all orthogonal to each other in nature, were not amenable to interpretation. Therefore an oblique rotation of the reference axes, called varimax rotation, was performed and factors derived. With schedule as the performance measuring criteria, a total of 6 success factors and 7 failure factors have emerged and when cost is the performance measuring criteria, a total of 7 success and failure factors have emerged. Similarly when quality is the performance criteria, a total of 7 success factors and 7 failure factors emerged while for dispute as the performance criteria 6 success factors and 6 failure factors emerged. Factor loadings <0.4 are suppressed in the analysis and those having loading values ≥ 0.4 are only taken for interpretations. The reliability of factor model was also checked with the communalities of each variable. Communalities of all the variables are found to be much greater than 0.3 that signifies that the factor model is reliable in the present study. The following paragraphs deal with the description of success and failure factors corresponding to individual performance measuring criteria.

3.6 Success Factors: Schedule Criterion

A total of six factors were extracted. These factors with their names representing their common and latent properties, the variance explained by each of them and the factor loadings of various attributes appearing in each factor are summarized in Table 3.7. The six factors collectively explain 79.88 % of the total variance. Names assigned to these factors are discussed in the following paragraphs.

3.6.1 Project Manager's Competence

As can be seen from Table 3.7, sixteen attributes with factor loadings ≥ 0.4 appear in this factor and it explains a variance of 27.95 %. Most attributes of this factor explain project manager's competence as key to success of the project. A competent manager has the *technical capability* and *monitoring capabilities*. He shows his trust in his project team by way of delegating the authority to his team. He *organises resources* through constant persuasion with his higher ups, he takes active part in *construction control meetings* held at site level, he acts as a catalyst

Table 3.7 Factor profile of project success attributes for schedule criterion (Reproduced 'with permission from ASCE')

Details of factor and the attributes	Factor loading
Factor_1 Project Manager's Competence (variance explained 27.95 %)	
Coordinating ability and rapport of PM with other contractors at site	0.898
Leadership quality of PM	0.870
Coordinating ability and rapport of PM with owner representatives	0.857
Authority to take day to day decisions by the PM's team at site	0.836
Regular budget update	0.766
Project manager's technical capability	0.757
Project manager's authority to take financial decision, selecting key team members, etc.	0.737
Construction control meetings	0.725
Availability of resources (funds, machinery, material, etc.) as planned throughout the project duration	0.691
Ability to delegate authority to various members of his team by PM	0.658
Training the human resources in the skill demanded by the project	0.657
Understanding of responsibilities by various project participants	0.618
Coordinating ability and rapport of PM with his team members and sub-contractor	0.569
Effective monitoring and feedback by PM	0.510
Developing and maintaining a short and informal line of communication among project team	0.431
Commitment of all parties to the project	0.417
Factor_2 Top Management Support (variance explained 15.86 %)	
Understanding operational difficulties by the owner engineer thereby taking appropriate decisions	0.833
Selection of PM with proven track record at an early stage by top management	0.817
Top management's backing up the plans and identify critical activities	0.781
Top management's enthusiastic support to the PM and project team at site	0.716
Developing and maintaining a short and informal line of communication among project team	0.625
Construction control meetings	0.605
Availability of resources (funds, machinery, material, etc.) as planned throughout the project duration	0.512
Factor_3 Monitoring, Feedback And Coordination (variance explained 15.24 %)	
Effective monitoring and feedback by the project team members	0.815
Positive attitude of PM, and project participants	0.746
Coordinating ability and rapport of PM with top management	0.735
Effective monitoring and feedback by PM	0.618
Understanding of responsibilities by various project participants	0.606
Coordinating ability and rapport of PM with his team members and sub-contractor	0.533
Top management's enthusiastic support to PM and project team at site	0.443
Factor_4 Favorable Working Condition (variance explained 8.72 %)	
Scope and nature of work well defined in the tender	0.752
Favorable social environment	0.689

(continued)

Table 3.7 (continued)

Details of factor and the attributes	Factor loading
Favorable climatic condition at the site	0.665
Monitoring and feedback by client	0.569
Factor_5 Commitment Of All Project Participants (variance explained 6.75 %)	
Favorable political and economic environment	0.706
Commitment of all parties to the project	0.679
Delegating authority to project manager by top management	0.617
Ability to delegate authority to various members of his team by PM	0.441
Factor_6 Owners Competence (variance explained 5.36 %)	
Timely decision by the owner or his engineer (reluctance or otherwise)	0.650
Monitoring and feedback by client	0.608
Training the human resources in the skill demanded by the project	−0.457

in training his human resources in the skill demanded by the project; he makes his people *committed* for the project through *effective leadership* and by acting in nonpartisan ways. All these attributes can be thought of originating from Project Manager's competence, hence the name.

3.6.2 Top Management Support

Most attributes of the second factor that explains 15.86 % variance (Table 3.7) indicate their common property to be supports of owners and top management. Owners support the project by *taking appropriate action* whenever the project faces operational difficulties. Top management extends support to the project by *selecting a project manager of proven track record at an early stage* and by *keeping short and informal line of communication.* Also a supportive top management *backs up the project plan* prepared by site management team. Taking part in *construction control meetings* and *making the resources available* for the planned duration also show support for the project.

3.6.3 Monitoring, Feedback and Coordination

Looking at the attributes under this factor it is difficult to give a common name. However, this factor has the attributes mainly focusing on *monitoring*, *feedback* and *coordination* among project participants. As can be seen from the section on literature review (Table 3.7), monitoring, feedback and coordination are given importance by most of the studies for the successful outcome of a project and identified as key factor responsible for success of many projects (Sayles and Chandler 1971; Pinto and Slevin 1988a). This factor explains 15.24 % variance.

3.6.4 Favorable Working Condition

This factor comprises of attributes like *scope and nature of work well defined* in the tender, *favorable social environment*, *favorable climatic condition at the site*, *monitoring and feedback by client*. This factor explains a variance of 8.72 %.

3.6.5 Commitment of All Project Participants

Project participants include internal and external participants. For the project to be successful not only internal factors are needed but external factors like political and economic environment are also needed. Delegating authority viz. top management delegating to project manager, which in turn delegates down to other team members also show the team commitment.

3.6.6 Owners Competence

Taking *timely decisions*, and regular *monitoring and feedback* of the progress of the project are some of the characteristics of a competent owner. The negative loading of attributes *training the human resources in the skill demanded by project* under this factor is bothersome and no meanings could be attached to this.

3.7 Failure Factors: Schedule Criterion

As discussed earlier the failure attributes were also subjected to factor analysis in the similar manner as that of success attributes. A total of seven factors emerged as failure factors that collectively explain 72.84 % of the total variance. The factors generated from the studies are described below. A summary of factors emerged, attributes associated with each of the factors, their factor loadings and variance explained by each factor is presented in Table 3.8.

3.7.1 Conflict Among Project Participants

This is the first failure factor containing seven attributes with factor loadings $\geq$0.4. The attributes under this factor mainly explains either the difference of opinion or lack of coherence in some way barring one or two attributes. This factor explains 15.34 % of variance.

Table 3.8 Factor profiles of project failure attributes for schedule criterion (reproduced with permission from ASCE)

Details of Factor and the attributes	Factor loading
Factor_1 Conflict Among Project Participants (variance explained 15.34 %)	
Tendency to pass on the blame to others	0.776
Conflicts between PM and other outside agency such as owner, sub-contractor, or other contractors	0.736
Conflicts between PM and top management	0.734
Poor human resource management and labor strike	0.645
Conflicts among team members	0.595
Project completion date specified but not yet planned by the owner	0.531
Mismatch in capabilities of client and architect	0.428
Factor_2 Project Manager's Ignorance (variance explained 11.34 %)	
Ignorance of appropriate planning tools and techniques by PM	0.825
Reluctance in timely decision by PM	0.772
Lack of understanding of operating procedure by the PM	0.674
Factor_3 Hostile Socio Economic Environment (variance explained 11.13 %)	
Hostile political and economic environment	0.938
Hostile social environment	0.928
Factor_4 Owner's Incompetence (variance explained 10.05 %)	
Inadequate project formulation in the beginning	0.790
Project completion date specified but not yet planned by the owner	0.612
Vested interest of client representative in not getting the project completed in time	0.524
Mismatch in capabilities of client and architect	0.494
Lack of understanding of operating procedure by the PM	0.400
Factor_5 Indecisiveness Of Project Participants (variance explained 9.98 %)	
Reluctance in timely decision by top management	0.823
Negative attitude of PM, and project participants	0.755
Holding key decisions in abeyance	0.493
Size and value of the project being large	−0.406
Factor_6 Harsh Climatic Condition At Site (variance explained 7.50 %)	
Harsh climatic condition at the site	0.777
Size and value of the project being large	0.598
Aggressive competition at tender stage	−0.581
Factor_7 Project Specific Factor And Aggressive Competition At Tender Stage (variance explained 7.50 %)	
Uniqueness of the project activities requiring high technical know-how	0.815
Aggressive competition at tender stage	0.453

3.7.2 *Project Manager's Ignorance*

The three attributes of this factor mainly indicate their common property as project manager's ignorance. *Ignorance of planning tools and lack of knowledge* will make project manager unable to identify and monitor the important activities that need to be carried out for schedule completion. He will not ask for timely help from top management due to unawareness of operating procedures resulting into time overrun for the project. This factor is accountable for 11.34 % of the variance explained. Contracting organizations are well advised not to compromise on the competence of project manager since an ignorant and less knowledgeable PM can spoil a project.

3.7.3 Hostile Socio-Economic Environment

The factor comprises of attributes like *hostile political and economic environment*; and *hostile social environment* and is accountable for 11.13 % of variance explained. During the follow up interviews with respondents the researchers came across a number of projects that had failed due to hostile socio-economic environment. Some projects couldn't even take off and some projects had to be shelved in the middle.

3.7.4 Owner's Incompetence

Four out of five attributes that have emerged in this factor emphasize owner related defaults leading to project failure. From the variables emerging it can be interpreted that an incompetent owner may go ahead with the project even with inadequately formulated project, he may freeze the completion date without proper planning and also may not be able to recruit competent person to look after his interests. All these result into unsuccessful outcome as far as schedule is concerned. This factor explains a variance of 10.05 %.

3.7.5 Indecisiveness of Project Participants

This factor explains a variance of 9.98 %. Indecisiveness in taking day to day decisions as well as holding key decisions have negative impact on schedule performance as seen in a number of projects mentioned by the respondents during follow up interviews. In one of the recently completed projects it was found that due to indecisiveness in furniture layout the entire lighting and building management system work was held up resulting in considerable loss of time.

3.7.6 Harsh Climatic Condition at Site

A section of respondents had worked under extreme climatic condition and could quote number of cases where the extreme conditions resulted into considerable loss of man-days. Harsh climatic condition not only has its heavy toll on the efficiency and productivity of the work force but it also causes difficulties in mobilizing the resources in time. If the size of the project is small these things can be managed however, when the size is large, time overrun is not at all ruled out. This factor explains 7.5 % variance.

3.7.7 Project Specific Factor and Aggressive Competition at Tender Stage

Attributes of this factor indicate project specific reasons as causes for failure. The uniqueness of the project activities may require participants to consume some initial time getting used to the project. This may result in loss of efficiency in the beginning that may have negative impact on the schedule. Aggressive competition at the tender stage might cause the lowest bidder to quote at very low margin that may not motivate him enough to work with full zeal and enthusiasm resulting into time overrun. This factor also explains 7.5 % variance.

3.8 Success Factors: Cost Criterion

The factor analysis is conducted separately on group of 30 success attributes and 23 failure attributes in order to extract the success and failure factors for cost criterion. The analysis extracted altogether seven success factors from the 30 success attributes and they explain a total of about 75 % of the variance. The details of the success factors are presented in Table 3.9 and they are described in the following paragraphs.

3.8.1 Project Manager's Competence

This is the first factor emerging in the cost criterion and it explains the largest variance of 22.15 %. The project manager's competence has four inherent aspects. The first is the combination of intrinsic traits that the project manager possesses such as his/her *technical capability, leadership quality, and positive attitude*. The second aspect is the empowerment of his team through *delegation of authority to take day to day decisions, making his/her team understand their responsibilities and generating a sense of commitment in them, developing and maintaining a short and informal line of communication among his/her team, and training the human resources in the skill demanded by the project*. The third aspect is to get empowered himself through *demanding authority to take financial decision, and selecting key team members*, etc., *and getting the required resources (funds, machinery, material*, etc.) *as planned throughout the project duration* from his/her higher ups. It is not enough to possess the skills mentioned above unless the project manager exerts himself/herself for the

Table 3.9 Factor profile of project success for cost criterion (Reprinted from Iyer and Jha 2005, with permission from Elsevier)

Details of factor and the attributes	Factor loading
Factor-1 Project Manager's Competence (variance explained 22.15 %)	
Authority to take day to day decisions by the PM's team at site	0.800
Construction control meetings	0.795
Regular budget update	0.795
Availability of resources (funds, machinery, material, etc.) as planned throughout the project duration	0.776
Project manager's authority to take financial decision, selecting key team members	0.749
Understanding of responsibilities by various project participants	0.740
Project manager's technical capability	0.701
Commitment of all parties to the project	0.665
Developing and maintaining a short and informal line of communication among project team	0.644
Training the human resources in the skill demanded by the project	0.632
Leadership quality of PM	0.574
Positive attitude of PM, and project participants	0.408
Factor_2 Top Management's Support (variance explained 11.41 %)	
Understanding operational difficulties by the owner engineer thereby taking appropriate decisions	0.786
Top management's enthusiastic support to the PM and project team at site	0.751
Top management's backing up the plans and identify critical activities	0.666
Delegating authority to project manager by top management	0.592
Selection of PM with proven track record at an early stage by top management	0.500
Timely decision by the owner or his engineer (reluctance or otherwise)	0.424
Developing and maintaining a short and informal line of communication among project team	0.407
Factor_3 Project Manager's Coordinating And Leadership Skill[a] (variance explained 10.28 %)	
Coordinating ability and rapport of PM with other contractors at site	0.880
Coordinating ability and rapport of PM with owner representatives	0.780
Training the human resources in the skill demanded by the project	0.515
Leadership quality of PM	0.505
Project manager's authority to take financial decision, selecting key team members.	0.453
Factor_4 Monitoring And Feedback (variance explained 9.28 %)	
Monitoring and feedback by top management	0.761
Timely decision by the owner or his engineer (reluctance or otherwise)	0.752
Selection of PM with proven track record at an early stage by top management	0.686
Favorable political and economic environment	0.672
Monitoring and feedback by client	0.576
Factor_5 Coordination Between Project Participants (variance explained 8.26 %)	
Coordinating ability and rapport of PM with top management	0.851
Coordinating ability and rapport of PM with his team members and sub-contractor	0.678
Effective monitoring and feedback by PM	0.451
Factor_6 Commitment Of Project Participants (variance explained 8.13 %)	
Ability to delegate authority to various members of his team by PM	0.639

(continued)

Table 3.9 (continued)

Details of factor and the attributes	Factor loading
Positive attitude of PM, and project participants	0.607
Effective monitoring and feedback by the project team members	0.572
Effective monitoring and feedback by PM	0.471
Commitment of all parties to the project	0.432
Factor_7 Owners Competence And Favorable Climatic Condition (variance explained 6.16 %)	
Favorable climatic condition at the site	0.832
Monitoring and feedback by client	0.655
Scope and nature of work well defined in the tender	0.580

[a] Taken with the first factor for subsequent discussion

project by getting involved in the project through *regular budget update and taking active part in construction control meetings*. All these four characteristics; inherent or personal traits; empowering team; getting empowered; and getting involved with the project are typical characteristics of a competent project manager. Several attributes emerging in this factor (Table 3.9) also address these characteristics only hence the name of the factor.

3.8.2 Top Management Support

In this study, the top management means both the contractors' and owners' top management. From the owners' side the attributes under this factor are: *understanding operational difficulties by the owner engineer thereby taking appropriate decisions*, and *timely decision by the owner or his engineer*. From the contractors' side the attributes are: *top management's enthusiastic support to the project manager (PM) and project team at site, top management's backing up the plans and identify critical activities, delegating authority to project manager by top management, selection of project manager with proven track record at an early stage by top management, and developing and maintaining a short and informal line of communication among project team*. These attributes are seen to emerge in the second factor (Table 3.9) and hence the name. It may be recalled that this factor has emerged as the second factor in the schedule criterion too.

3.8.3 Project Manager's Coordinating and Leadership Skill

The third factor shares three common attributes with the first factor out of five attributes that emerged. Although the first two attributes (Table 3.9) predominantly describe the coordinating and leadership skill of a project manager, they cannot be taken as different from the first factor which covers the overall aspects of project manager's traits or competence. Hence this factor along with the first factor can be said to explain the common latent property and together explain 33.43 % variance (22.15 % from factor 1 and 10.28 % from factor 3).

3.8.4 Monitoring and Feedback

The attributes under this factor mainly focus on monitoring and feedback by the project participants. The first and last attributes under this factor directly mention this whereas second and third attribute under this factor indicate action by a watchful owner and top management, which is possible only when these participants are monitoring the projects. As can be seen from the literature review chapter, monitoring and feedback is given importance by most of the studies for the successful outcome of a project and identified as key factor responsible for success of many projects. This factor explains a variance of 9.28 %.

3.8.5 Coordination Between Project Participants

The high loading attributes here are coordinating ability and rapport of project manager with top management, team members, and subcontractors. Effective monitoring and feedback by a project manager also makes it in this factor. These attributes mainly point toward the interaction or personal rapport with the participants. Many a time personal rapport with different project participant can save a lot of cost. Individual rapport can lead to each other providing helping hand and thereby sometimes reducing the necessary paper work or action and paper work going simultaneously, thereby reducing considerable time and hence cost finally.

3.8.6 Commitment of Project Participants

The positive attitude as well as monitoring and feedback by project manager and the participants show the commitment. Project manager also shows his/her commitment by delegating the authority to other members of the team and not sticking with all the powers granted to him by the top management. This way a project manager ensures commitment.

3.8.7 *Owner's Competence and Favorable Climatic Condition*

This factor explains the lowest variance (6.16 %) among all the factors and indicates the two broad aspects: owner's competence and climatic conditions. A competent owner would have its scope of work well outlined and presented to the contractor and he/she would closely monitor his/her project regarding its progress, budget, quality, and other aspects. Providing favorable climatic condition is beyond the control of an owner or a contractor until one is given the choice to select his project location. As this factor explains two diverse aspects, the name of this factor has been kept as owner's competence and favorable climatic condition. Chan et al. (2001b) have also recognized an owner's competence as the most important factor for Design and Build (D&B) Projects.

3.9 Failure Factors: Cost Criterion

As mentioned in the previous section, the factor analysis was separately performed on the 23 failure attributes also to understand the intrinsic characteristics of failure attributes corresponding to cost criterion. This resulted in the extraction of seven failure factors and they explain about 70 % of the variance. The details of the failure factors are presented in Table 3.10 and are described below.

3.9.1 *Conflict Among Project Participants*

As can be seen from Table 3.10 this factor explains 19.95 % of the variance, the highest of all factors and this contains 11 attributes with high factor loadings ($\geq$0.4). The majority of the attributes barring one or two attributes under this factor mainly explain either the difference of opinion or the lack of coherence in some way. This factor explains 19.95 % of variance. It is expected of the top management to devise various measures to tackle this issue in the organization.

3.9.2 *Ignorance and Lack of Knowledge of Project Manager*

The attributes having high loading in this factor are: ignorance of appropriate planning tools and techniques by a project manager; reluctance in timely decision by the project manager; lack of understanding of operating procedure by the project manager; and conflicts among team members. This factor accounts for 10.78 % of variances explained. This factor is in line with one of the established

Table 3.10 Factor profile of project failure attributes for cost criterion (Reprinted from Iyer and Jha 2005, with permission from Elsevier)

Details of factor and the attributes	Factor loading
Factor_1 Conflict Among Project Participants (variance explained 19.95 %)	
Poor human resource management and labour strike	0.779
Mismatch in capabilities of client and architect	0.752
Negative attitude of PM, and project participants	0.720
Vested interest of client representative in not getting the project completed in time	0.694
Project completion date specified but not yet planned by the owner	0.639
Conflicts among team members	0.617
Conflicts between PM and top management	0.595
Conflicts between PM and other outside agency such as owner, sub-contractor or other contractors	0.541
Lack of understanding of operating procedure by the PM	0.486
Tendency to pass on the blame to others	0.470
Reluctance in timely decision by top management	0.443
Factor_2 Ignorance And Lack Of Knowledge (variance explained 10.79 %)	
Ignorance of appropriate planning tools and techniques by PM	0.839
Reluctance in timely decision by PM	0.746
Lack of understanding of operating procedure by the PM	0.650
Conflicts among team members	0.402
Factor_3 Project Specific Factor (variance explained 9.55 %)	
Inadequate project formulation in the beginning	0.786
Conflicts between PM and other outside agency such as owner, sub-contractor or other contractors	0.604
Tendency to pass on the blame to others	0.581
Conflicts between PM and top management	0.535
Holding key decisions in abeyance	0.445
Uniqueness of the project activities requiring high technical know-how	0.428
Factor_4 Hostile Socio Economic And Climatic Condition (variance explained 8.25 %)	
Hostile political and economic environment	0.894
Hostile social environment	0.591
Harsh climatic condition at the site	0.440
Factor_5 Reluctance in timely decision (variance explained 8.17 %)	
Reluctance in timely decision by top management	0.704
Size and value of the project being large	−0.631
Presence of crisis management skill of PM	−0.630
Factor_6 Aggressive Competition At Tender Stage (variance explained 6.67 %)	
Aggressive competition at tender stage	0.824
Harsh climatic condition at the site	−0.420
Holding key decisions in abeyance	−0.540
Factor_7 Short Bid Preparation Time (variance explained 5.65 %)	
Urgency emphasized by the owner while issuing tender	0.783

facts in literature and the findings of this study given in previous section. While a competent project manager becomes responsible for the success of a project, ignorance and lack of knowledge of project manager can cause failure as seen from this factor. Contracting organizations are well advised not to compromise on the competence of project manager. The top management can devise means to supplement the knowledge needs of project participants by providing training at regular intervals.

3.9.3 Project Specific Factors

This factor points to two broad categories of attributes. The first category has project specific attributes viz. inadequate project formulation in the beginning; uniqueness of the project activities requiring high technical know-how; holding key decisions in abeyance and the other category has attributes related to nonexistence of cooperation among project participants in the form of conflicts and passing blame. Accordingly, the name of the factor has been suggested. The factor accounts for 9.55 % variance explanation.

3.9.4 Hostile Socio Economic and Climatic Condition

This factor affects the cost performance adversely in the form of frequent stoppage of work, labour unrest, and reduced productivity. Respondents have narrated many projects like the famous Enron power project (India) and other projects where political views against the project have led either to inflate the schedule/cost manifold or has led to shelving of the project itself. Respondents also mention the current status of Tehri dam project and Sardar Sarovar Project (India), which have also suffered on account of opposition by a section of people resulting into severe cost and time overrun. These two cases are internationally known for hostile socio economic condition.

3.9.5 Reluctance in Timely Decision

The name of this factor is evident, as it has been directly taken from the only attribute under this factor that has positive factor loading. No meanings could be assigned to parameters with negative factor loading nor could this factor be clubbed with any other failure factors explained in this section. The reluctance in decision making could be on account of many reasons for example the possible repercussion if something goes wrong. This is an important and evident factor, which are generally talked by professionals based on their experience.

3.9.6 Aggressive Competition at Tender Stage

Although aggressive competition at tender stage should enhance the chances for improving the cost performance of the project however, it is not true in practice. Since most of the times such projects land up in disputes arising out of petty things and claims/counterclaims extend the duration of the project resulting into large cost overrun. This is why probably respondents rate this as failure factor rather than a success factor.

3.9.7 Short Bid Preparation Time

The project duration generally includes entire duration right from conception/ approval stage till execution and handing over. In order to gain time for execution or unforeseen events, owners or their representatives tend to squeeze the bid preparation time itself. In an attempt to get the job contractors are unable to force the owners to provide a reasonable time to quote for the project. Ideally reasonable time should be allowed for proper site investigation, and collection of relevant details required for estimation purpose etc. The short bid preparation time leads to a number of errors/omissions on contractor's part which they try to settle later through claims. This raises disputes and finally the project lands up with schedule and cost overruns. This factor with one attribute alone explains a variance of 5.65 %.

3.10 Success Factors: Quality Criterion

The details of the seven factors, the factor loading, and variance explained by each factor are given in Table 3.11. This table also gives the details of the attributes emerged under each factor. The individual factors are described below.

3.10.1 Project Manager's Competence

The attributes emerging under this factor indicate the importance of a competent project manager similar to other performance criterion. This factor explains a variance of 19.52 %. Sometimes the mere involvement of project manager in site activities can lift the morale of his team members and they start working with full zeal and enthusiasm to achieve the desired quality level. A competent project manager can contribute by bringing in innovative solutions and guiding his/her team members in performing their tasks to the desired quality level.

Table 3.11 Factor profile of project success attributes for quality criterion (Reprinted from Jha and Iyer 2006, with permission from Taylor and Francis)

Details of factor and the attributes	Factor loading
Factor_1 Project Manager's Competence (variance explained 19.52 %)	
Coordinating ability and rapport of PM with owner representatives	0.82
Authority to take day to day decisions by the PM's team at site	0.77
Commitment of all parties to the project	0.75
Understanding of responsibilities by various project participants	0.71
Project manager's authority to take financial decision, selecting key team members	0.71
Coordinating ability and rapport of PM with other contractors at site	0.68
Project manager's technical capability	0.66
Scope and nature of work well defined in the tender	0.50
Positive attitude of PM, and project participants	0.43
Training the human resources in the skill demanded by the project	0.43
Construction control meetings	0.50
Ability to delegate authority to various members of his team by PM	0.44
Leadership quality of PM	0.40
Factor_2 Top Management's Support (variance explained 15.94 %)	
Top management's backing up the plans and identify critical activities	0.83
Top management's enthusiastic support to the PM and project team at site	0.83
Understanding operational difficulties by the owner engineer thereby taking appropriate decisions	0.83
Availability of resources (funds, machinery, material, etc.) as planned throughout the project duration	0.61
Positive attitude of PM, and project participants	0.60
Training the human resources in the skill demanded by the project	0.49
Developing and maintaining a short and informal line of communication among project team	0.45
Timely decision by the owner or his engineer (reluctance or otherwise)	0.47
Effective monitoring and feedback by the project team members	0.50
Factor_3 Top Management's Competence (variance explained 9.91 %)	
Selection of PM with proven track record at an early stage by top management	0.87
Delegating authority to project manager by top management	0.86
Developing and maintaining a short and informal line of communication among project team	0.52
Construction control meetings	0.51
Factor_4 Interaction Between Project Participants-Internal (variance explained 9.62 %)	
Coordinating ability and rapport of PM with top management	0.84
Coordinating ability and rapport of PM with his team members and sub-contractor	0.82
Positive attitude of PM, and project participants	0.47
Developing and maintaining a short and informal line of communication among project team	0.42
Construction control meetings	0.41
Factor_5 Owners Competence (variance explained 9.32 %)	
Monitoring and feedback by client	0.81
Monitoring and feedback by top management	0.79

(continued)

Table 3.11 (continued)

Details of factor and the attributes	Factor loading
Timely decision by the owner or his engineer (reluctance or otherwise)	0.61
Ability to delegate authority to various members of his team by PM	0.54
Factor_6 Monitoring And Feedback By Project Participants (variance explained 8.16 %)	
Effective monitoring and feedback by PM	0.90
Effective monitoring and feedback by the project team members	0.64
Favorable climatic condition at the site	0.58
Commitment of all parties to the project	0.43
Factor_7 Interaction Between Project Participants-External (variance explained 4.56 %)	
Leadership quality of PM	0.57
Coordinating ability and rapport of PM with other contractors at site	0.41
Training the human resources in the skill demanded by the project	−0.41

3.10.2 Top Management Support

The attributes emerging under factor_2 and factor_3 accounting for a combined variance of about 25 % (15.94 + 9.91 %) explain the top management support and their competence. As can be seen the top management support is essential for achieving desired quality on account of mainly four issues: It is the top management's prerogative to set all the policy issues (including quality policy) and control resources. Also they arrange training of human resources involved in the project and they have big role to play in identifying the project manager for the project. It can be seen that the top management controls all the key factors and hence their support is highly desired for the quality compliance. With amount of variance of 25 % explained through two factors top management support practically the most important of all factors.

3.10.3 Interaction Between Project Participants

This factor is made up of Factor_4 and Factor_7. Any project involves interaction between different project participants (stakeholders). The participants include the internal participants like contractor's team member as well as the external team members like different subcontractors and vendors. Most of the activities require proper understanding of the needs of the others. There are instances when the quality of the project suffers for want of proper interaction between the participants. This fact is more vivid if one executes projects, which involve multiple categories of work say for example: civil works; electrical works; mechanical works; HVAC (heating, ventilation, and air-conditioning); and building automation etc. One can appreciate the havoc created to the quality of project activities on account of lack of interaction among project participants. The coordinating ability and positive attitude of project participants are great assets in such conditions. A short and informal line of communication as well as regular construction control meetings among project team further supports achievement of desired quality level. In the literature there are numerous examples where a number of difficulties have been faced due to lack of interaction between project participants.

3.10.4 Owners Competence

The owners play an important role in achieving the desired quality level. Not only they are responsible for the preparation of a clear and unambiguous specification they must also monitor the actual work at site. It is now recognized that for the clients' inspectors to work with the contractor to establish good quality control procedures before the work is done, is much more effective than walking around (Barnes 1987). Also if any case of any discrepancies or deviation from the specification is observed it should be communicated immediately to the concerned person. If the owners desire quality job, they should stick to the specification since any relaxation in quality performance even for few times can set bad precedence.

3.10.5 Monitoring and Feedback

Mainly two attributes emerging in this factor explain predominantly aspects of monitoring and feedback. Since the other attributes appear as a strange combination, the factor is named after monitoring and feedback. This is ranked sixth based on the variance 8.16 % (out of 77.03 % of total variance) explained. Proper monitoring and timely feedback helps in controlling the workmanship and it enhances the quality of a project. If each of the activity of a project is monitored effectively and instances of poor workmanship, and improper usage of material, labor or plant and machinery are reported promptly it aids in achieving the desired quality level. Committed participants would come forward in sticking to the quality plan and they would follow the accepted technical practices to carry out the different project activity.

3.11 Failure Factors: Quality Criterion

Factor analyses of responses on 22 failure attributes on quality criterion resulted in seven failure factors and they account for about 70 % of the variance explained. The individual factors, the attributes emerging under each factor, the factor loading, and the variance explained by individual factors are given in Table 3.12. The failure factors are discussed in the following paragraphs.

Table 3.12 Factor profile of project failure attributes for quality criterion (Reprinted from Jha and Iyer 2006, with permission from Taylor and Francis)

Details of factor and the attributes	Factor loading
Factor_1 Conflict Among Project Participants (variance explained 16.32 %)	
Negative attitude of PM, and project participants	0.75
Poor human resource management and labour strike	0.74
Mismatch in capabilities of client and architect	0.71
Vested interest of client representative in not getting the project completed in time	0.66
Holding key decisions in abeyance	0.63
Conflicts among team members	0.51
Reluctance in timely decision by top management	0.42
Conflicts between PM and top management	0.41
Factor_2 Hostile Socio Economic And Climatic Condition (variance explained 12.57 %)	
Hostile social environment	0.83
Hostile political and economic environment	0.75
Harsh climatic condition at the site	0.59
Inadequate project formulation in the beginning	0.55
Urgency emphasized by the owner while issuing tender	0.48
Factor_3 Ignorance And Lack Of Knowledge (variance explained 11.66 %)	
Ignorance of appropriate planning tools and techniques by PM	0.78
Reluctance in timely decision by PM	0.77
Lack of understanding of operating procedure by the PM	0.75
Reluctance in timely decision by top management	0.41
Factor_4 Faulty Project Conceptualization (variance explained 7.93 %)	
Tendency to pass on the blame to others	0.78
Project completion date specified but not yet planned by the owner	0.59
Conflicts between PM and top management	0.44
Factor_5 Project Specific Factors (variance explained 7.83 %)	
Uniqueness of the project activities requiring high technical know-how	0.87
Size and value of the project being large	0.74
Factor_6 Conflict among project participants (variance explained 7.21 %)	
Conflicts between PM and other outside agency such as owner, sub-contractor, or other contractors	0.78
Conflicts between PM and top management	0.41
Conflicts among team members	0.41
Size and value of the project being large	−0.45
Factor_7 Aggressive Competition At Tender Stage (variance explained 6.80 %)	
Aggressive competition at tender stage	0.82
Urgency emphasized by the owner while issuing tender	0.48

3.11.1 Conflict Among Project Participants

Looking at Table 3.12, it can be observed that majority of the attributes contained in Factor_1, and Factor_6 barring few attributes either represent difference in opinions or lack of coherence in some way or other. It indicates that all blue and white-collar workers must work in unison otherwise it leads to improper quality level achieved at site. Also the organization where people at higher hierarchy tends to pass the blame to lower hierarchy people, the achievement of desired quality always remains in doubt. As discussed earlier recognizing quality as a team work the management should create suitable environment to build up a team by plugging all such causes giving rise to adversarial relationship among team members.

3.11.2 Hostile Socio Economic and Climatic Condition

The hostile work environment affects the quality of a construction adversely as suggested by this analysis. This has also been observed by other industries. Poor work environment not only decreases productivity but it also affects the project quality. Also harsh climatic conditions give rise to fatigued work force leading to poor quality. As can be seen from Table 3.12 this factor accounts for 12.57 % of variance explained.

3.11.3 Ignorance and Lack of Knowledge

If project participants lack in job knowledge and they ignore the appropriate planning tools and established quality norms it results into poor quality. This factor is accountable for 11.66 % of variance explained. The top management should devise means to supplement the knowledge needs of project participants by providing training at regular intervals. Proper recruiting policy and arranging in-house training program for the project team members can also tackle these aspects.

3.11.4 Faulty Project Conceptualization

Attributes under this factor represent faulty project conceptualization and conflict between a project manager and top management. However, the name of the factor has been kept, as 'faulty project conceptualization' since the conflict among participants has predominantly been present in Factor_1 and Factor_6. If the project completion date has been frozen without arranging inputs and proper planning, they lead to hasty and unsystematic work toward the end of the project resulting into project quality taking a backseat. All these haste leads to relaxation in quality specification from owners' side too, as they tend to overlook the deviation by the contractor from the agreed technical specification. The contractor on his part tries to save on time by adopting shortcuts and bad technical practices. All these lead to poor quality.

3.11.5 Project Specific Factors

As can be seen the attributes emerging under this factor are project specific. The emergence of the attribute 'uniqueness of the project activities requiring high technical know-how' in the failure attribute indicates that if a project involves certain unique activities that the project people may not have executed on previous projects it contributes negatively in achieving desired quality. Some learning time may also be required for the people involved with these activities. Apart from this attribute, if the size and contract value of the project is large the limited project people may not be able to do justice in all areas and this may adversely affect the project quality.

3.11.6 Aggressive Competition at Tender Stage

Aggressive competition sometimes forces the bidders to quote low for the project. Once awarded the project they are not motivated enough to do a quality job. To make some profit out of the project they sometime try to use inferior materials and bad technical practices leading to poor quality. The problem of low bid is quiet common in case of government owned projects. While it is perfectly logical for the government to accept low bids being the guardian of public funds, selection of low bidder more often causes problem to the project than doing any good to the project. Also the low bidder sometimes resort to subcontracting the entire project to unqualified contractors leading to poor quality.

3.12 Success Factors: No-Dispute Criterion

Like the previous analyses the responses to the 28 success attributes on the no-dispute criterion are also subjected to factor analysis. The summary of the analysis results are given in Table 3.13 and the emerged factors are discussed in the following paragraphs.

3.12.1 Project Manager's Competence and Top Management Support

Even in this criterion, the project manager's competence and top management support have emerged out to be the most important success factors explaining 28.38 and 20.10 % variances out of total variance of 76.08 % explained by the factor analysis. It is true that if project manager and the top management are competent and supportive to each other, even the major disputes can be handled in an amicable manner.

3.12.2 Owner's Competence

This is the third in the order of importance gaged from the variance explained by the individual factor. Owner's competence can be gaged from attributes such as the monitoring and feedback, the clarity of scope and nature of work, and the timely decision making capabilities. The potential for disputes avoidance is the greatest in the design and contract preparation phase of a project, when problems can be rectified with fewer cost and time implications than at any other time in the construction process. For this reason, competence of owner is very critical. Thompson (1991) views a client's role during project development and implementation as crucial to the success of the project. Project managers, however, often lack support from the client organization's top management. Some of the attributes appearing under this factor explain project manager's competence and they have been dealt under the factor 'project manager's competence'. The other factors that are considered to pave path for reduction of disputes are *favorable political and social environment*; *availability of trained resources*; and *regular budget update*.

3.13 Failure Factors: No-Dispute Criterion

Summary of the factor analysis results of 22 failure attributes corresponding to no-dispute criterion given in Table 3.14 indicates that the analysis has explained 68.14 % variances. The top factor of course has been *conflict among project participants*. One can guess this factor to the most critical based on his intellectual wisdom. However, this factor's emergence in the analysis strengthens the common belief.

The other factors as can be seen in Table 3.14 are *ignorance and lack of knowledge of project manager*; *hostile socio-economic environment*; *indecisiveness of project participants and negative attitude*; *faulty project conceptualization*; and *harsh climatic condition*.

It can be observed from Tables 3.7, 3.8, 3.9, 3.10, 3.11, 3.12, 3.13 3.14 that there are a number of common factors across all four-performance criteria having identical attributes emerging in them, while some other factors emerged predominantly in only one or more performance criteria. Taking the union of all common and uncommon factors across all four-performance criteria, there are thus 11 success factors and 9 failure factors. These factors are summarized in Table 3.15. The success and failure factors in this table are arranged in decreasing order of variance explained in one criterion or other.

Table 3.13 Factor structure for project success attributes for no-dispute criterion

Details of factor and the attributes	Loading
Factor_1 Project Manager's Competence (variance explained 28.38 %)	
Effective monitoring and feedback by PM	0.861
Leadership quality of PM	0.844
Effective monitoring and feedback by the project team members	0.842
Coordinating ability and rapport of PM with other contractors at site	0.817
Understanding of responsibilities by various project participants	0.808
Coordinating ability and rapport of PM with owner representatives	0.806
Coordinating ability and rapport of PM with his team members and sub-contractor	0.737
Authority to take day to day decisions by the PM's team at site	0.735
Project manager's authority to take financial decision, selecting key team members, etc.	0.704
Project manager's technical capability	0.684
Regular budget update	0.637
Coordinating ability and rapport of PM with top management	0.537
Top management's backing up the plans and identify critical activities	0.464
Factor_2 Top Management's Support (variance explained 20.10 %)	
Top management's enthusiastic support to the project manager (PM) and project team at site	0.838
Delegating authority to project manager by top management	0.823
Top management's backing up the plans and identify critical activities	0.814
Understanding operational difficulties by the owner engineer thereby taking appropriate decisions	0.812
Selection of PM with proven track record at an early stage by top management	0.808
Construction control meetings	0.761
Timely decision by the owner or his engineer (reluctance or otherwise)	0.549
Availability of resources (funds, machinery, material, etc.) as planned throughout the project duration	0.483
Scope and nature of work well defined in the tender	0.412
Factor_3 Owners Competence (variance explained 9.23 %)	
Positive attitude of PM, and project participants	0.763
Monitoring and feedback by client	0.741
Scope and nature of work well defined in the tender	0.706
Timely decision by the owner or his engineer (reluctance or otherwise)	0.482
Coordinating ability and rapport of PM with top management	0.466
Coordinating ability and rapport of PM with his team members and sub-contractor	0.434
Factor_4 Favorable Political And Economic Environment (variance explained 7.71 %)	
Favorable political and economic environment	0.839
Commitment of all parties to the project	0.830
Monitoring and feedback by top management	0.607
Factor_5 Availability Of Trained Resources (variance explained 6.80 %)	
Training the human resources in the skill demanded by the project	0.793
Developing and maintaining a short and informal line of communication among project team	0.697
Availability of resources (funds, machinery, material, etc.) as planned throughout the project duration	0.662
Factor_6 Regular budget update (variance explained 3.86 %)	
Regular budget update	0.533

Table 3.14 Factor structure of project failure attributes for no-dispute criterion

Factor structure	Loading
Factor_1 Conflict Among Project Participants (variance explained 17.64 %)	
Conflicts between PM and other outside agency such as owner, sub-contractor or other contractors	0.812
Tendency to pass on the blame to others	0.793
Conflicts among team members	0.671
Project completion date specified but not yet planned by the owner	0.617
Conflicts between PM and top management	0.596
Lack of understanding of operating procedure by the PM	0.503
Vested interest of client representative in not getting the project completed in time	0.476
Ignorance of appropriate planning tools and techniques by PM	0.465
Poor human resource management and labour strike	0.438
Mismatch in capabilities of client and architect	0.432
Negative attitude of PM, and project participants	0.416
Factor_2 Ignorance And Lack Of Knowledge (variance explained 14.90 %)	
Lack of understanding of operating procedure by the PM	0.613
Vested interest of client representative in not getting the project completed in time	0.433
Ignorance of appropriate planning tools and techniques by PM	0.668
Poor human resource management and labor strike	0.697
Reluctance in timely decision by PM	0.815
Size and value of the project being large	−0.648
Factor_3 Hostile Socio Economic Environment (variance explained 10.94 %)	
Hostile political and economic environment	0.746
Uniqueness of the project activities requiring high technical know-how	0.744
Hostile social environment	0.740
Conflicts between PM and top management	0.430
Factor_4 Indecisiveness And Negative Attitude (variance explained 8.58 %)	
Holding key decisions in abeyance	0.804
Mismatch in capabilities of client and architect	0.564
Negative attitude of PM, and project participants	0.586
Factor_5 Faulty Project Conceptualization (variance explained 8.39 %)	
Inadequate project formulation in the beginning	0.761
Urgency emphasized by the owner while issuing tender	0.611
Aggressive competition at tender stage	−0.477
Factor_6 Harsh Climatic Condition (variance explained 7.69 %)	
Harsh climatic condition at the site	0.888
Reluctance in timely decision by top management	0.624

It can also be observed that only three success factors: Project manager's competence (F_1), Top management support (F_2), and Owners competence (F_6) are common across all the four performance criteria. While F_1 and F_2 are appearing as Factor_1 and Factor_2 across all the performance criteria, F_6 has appeared as Factor_6, Factor_7, Factor_5, and Factor_3 corresponding to schedule, cost, quality, and no-dispute performance criteria respectively. The balance eight-success factors out of a total of the eleven factors are also shown appearing in this table at different numbers in different performance criterion. It is also worth noting

Table 3.15 Pooled list of factors identified under various performance evaluation parameters (Reprinted from Iyer and Jha 2006, with permission from ASCE)

S. No	Factor type	Factor names	Factor identification number after pooling	Factors numbers as obtained after factor analysis in schedule performance criteria	Factors numbers as obtained after factor analysis in cost performance criteria	Factors numbers as obtained after factor analysis in quality performance criteria	Factors numbers as obtained after factor analysis in no-dispute performance criteria
1	Success factors	Project manager's competence	F_1	Factor_1	Factor_1 and Factor_3	Factor_1	Factor_1
2		Top management support	F_2	Factor_2	Factor_2	Factor_2 and Factor_3	Factor_2
3		Monitoring and feedback	F_3	Factor_3	Factor_4	Factor_6	–
4		Favorable working condition	F_4	Factor_4	Factor_7	–	Factor_4
5		Commitment of all project participants	F_5	Factor_5	Factor_6	–	–
6		Owners competence	F_6	Factor_6	Factor_7	Factor_5	Factor_3
7		Interaction between project participants-internal	F_7	–	–	Factor_4	–
8		Interaction between project participants-external	F_8	–	–	Factor_7	–
9		Good coordination among project participants	F_9	Factor_3	Factor_5	–	–
10		Availability of trained resources	F_{10}	–	–	–	Factor_5
11		Regular budget update	F_{11}	–	–	–	Factor_6

(continued)

Table 3.15 (continued)

S. No	Factor type	Factor names	Factor identification number after pooling	Factors numbers as obtained after factor analysis in schedule performance criteria	Factors numbers as obtained after factor analysis in cost performance criteria	Factors numbers as obtained after factor analysis in quality performance criteria	Factors numbers as obtained after factor analysis in no-dispute performance criteria
12	Failure factors	Conflict among project participant	F_{12}	Factor_1	Factor_1	Factor_1and Factor_6	Factor_1
13		Project manager's ignorance and lack of knowledge	F_{13}	Factor_2	Factor_2	Factor_3	Factor_2
14		Hostile socio economic environment	F_{14}	Factor_3	Factor_4	Factor_2	Factor_3
15		Owner's incompetence	F_{15}	Factor_4	–	–	–
16		Indecisiveness of project participants	F_{16}	Factor_5	Factor_5	–	Factor_4
17		Harsh climatic condition at site	F_{17}	Factor_6	Factor_4	Factor_2	Factor_6
18		Aggressive competition during tender stage	F_{18}	Factor_7	Factor_6	Factor_7	–
19		Negative attitude of project participants	F_{19}	–	–	–	Factor_4
20		Faulty project conceptualization	F_{20}	–	Factor_7	Factor_4	Factor_5

here that, Factor_3, a success factor in schedule criterion is found emerging as two distinct factors: "monitoring and feedback" and "coordination" and accordingly they are shown at two locations corresponding to F_3 and F_9 under schedule criterion.

On the other hand, attributes appearing in Factor_1 and Factor_3 in cost criterion (Table 3.9) inherently explain the same properties that of 'project manager's competence' and hence both factors are clubbed to represent one factor F_1 in cost criterion. Similarly the attributes appearing in Factor_2 and Factor_3 in quality criterion explain 'top management's support' and hence both of them are pooled in F_2 under quality criterion.

From the failure factors appearing in Table 3.15, it can be observed that four factors F_{12}, F_{13}, F_{14} and F_{17} are common across all the four performance criteria. While F_{12} is the first factor across all the four criteria, F_{13}, and F_{14} appear at next two levels. The balance five-failure factors also appear at different levels in this table in different performance criteria. In schedule and cost criteria *indecisiveness of project participants* has appeared to be a dominant failure factor while in no-dispute criterion this factor has appeared along with *negative attitude of project participants*. Since the influence of these two would vary differently for different criteria, they are therefore split and taken as two items as F_{16} and F_{19}.

It is also established in the statistics that variance explained by factors may not be a measure of intensity or importance of the factor in any performance criterion (Iyer 1996). The next objective of the study is thus set to understand the criticality of these factors on different project performance measuring criterion as well as on an overall basis. Therefore the second stage questionnaire using these 20 factors as explanatory variables and contribution of these factors (variables) on actual performance of the choice projects as response variable is developed and responses sought as explained below.

3.14 Summary and Conclusions

The attributes affecting the project performance objectives of schedule, cost, quality, and no-dispute compliances are discussed in this chapter. A total of 55 such attributes were identified. These attributes were segregated in three groups based on the mean values. The first group of attributes were considered to contribute positively in achieving the stated performance objectives, the second group of attributes were considered neutral causing neither positive effect nor negative effect in the performance objectives, and the third group of attributes affected adversely. They are referred to as success, neutral and failure attributes. The number of attributes appearing in the first group were 31, 2 in second group, and the remaining 22 attributes in the third group. Only two groups of attributes, success and failure attributes were taken for further study.

In order to understand the success and failure attributes in a better way and to reduce the number of attributes, the attributes were subjected to factor analyses separately for the success and failure groups of attributes.

Factors extracted under the four different performance evaluation parameters (schedule, cost, quality, and no-dispute) were found to be different hence a union of all factors is considered for further study. Thus a total of 20 factors (including 11 success factors and 9 failure factors) were identified and second level questionnaire survey was developed using these factors.

References

Andersen, E. S., Birchall, D., Jessen, S. A., & Money, A. H. (2006). Exploring project success. *Baltic Journal of Management, 1*(2), 127–147.

Ashley, D., Jaselskis, E., & Lurie, C. B. (1987). The determinants of construction project success. *Project Management Journal, 18*(2), 69–79.

Avots, I. (1969). Why does project management fail? *California Management Review, 12*(1), 77–82.

Baker, B. N., Murphy, D. C., & Fisher, D. (1983). *Factors affecting project success. Project Management Handbook* (pp. 669–685). New York: Van Nostrand Reinhold.

Barnes, M. (1987). Construction project management. Seminar on construction project management, 21–22 Oct, London, England.

Bonnal, P., Gourc, D., & Lacoste, G. (2002). The life cycle of technical projects. *Project Management Journal, 33*(1), 12–19.

Chan, A. P. C., Ho, D. C. K., & Tam, C. M. (2001a). Effect of interorganizational teamwork on project outcome. *Journal of Management in Engineering, 17*(1), 34–40.

Chan, A. P. C., Ho, D. C. K., & Tam, C. M. (2001b). Design and build project success factors: Multivariate analysis. *Journal of Construction Engineering and Management, ASCE, 127*(2), 93–100.

Chan, D. W. M., & Kumaraswamy, M. M. (1997). A comparative study of causes of time overruns in Hong Kong construction projects. *International Journal of Project Management, 15*(1), 55–63.

Chitkara, K. K. (1998). *Construction project management: Planning, scheduling and controlling*. New Delhi: McGraw Hill.

Chua, D. K. H., Kog, Y. C., & Loh, P. K. (1999). Critical success factors for different project objectives. *Journal of Construction Engineering and Management, ASCE, 125*(3), 142–150.

Chua, D. K. H., Kog, Y. C., Loh, P. K., & Jaselskis, E. J. (1997). Model for construction budget performance—neural network approach. *Journal of Construction Engineering and Management, ASCE, 123*(3), 214–222.

Cleland, D. I., & King, W. R. (1983). *Systems analysis and project management*. New York: McGraw Hill.

Dumont, P. R., Gibson, G. E., & Fish, J. R. (1997). Scope management using project definition rating index. *Journal of Management in Engineering, ASCE, 13*(5), 54–60.

Hughes, M. W. (1986). Why projects fail: The effects of ignoring the obvious. *Industrial Engineering, 18*, 14–18.

Iyer, K. C., & Jha, K. N. (2005). Factors affecting cost performance: Evidence from Indian construction projects. *International Journal of Project Management, 23*(4), 283–295.

Iyer, K. C. (1996). Identification and evaluation of dispute prone clauses in Indian construction contracts. A *Doctoral thesis* submitted at Indian Institute of Technology, Madras.

Jha, K. N., & Iyer, K. C. (2006). Critical determinants of project coordination, *International Journal of Project Management*, *24*(4), 314–322.

Jha, K. N., & Iyer, K. C. (2007). Commitment, coordination, competence and the iron triangle. *International Journal of Project Management, 25*(5), 527–540.

Lam, E. W. M., Chan, A. P. C., & Chan, D. W. M. (2008). Determinants of successful design-build projects. *Journal of Construction Engineering and Management, 134*(5), 333–341.

Locke, D. (1984). *Project management*. New York: St Martins Press.

Martin, C. C. (1976). *Project Management*. New York: Amaco.

Mansfield, N. R., & Odeh, N. S. (1991). Issues affecting motivation on construction projects. *International Journal of Project Management, 9*(2), 93–98.

Mansfield, N. R., Ugwu, O. O., & Doran, T. (1994). Causes of delay and cost overruns in Nigerian construction projects. *International Journal of Project Management, 12*(4), 254–260.

Mcneil, H. J., & Hartley, K. O. (1986). Project planning and performance. *Project Management Journal, 17*(1), 36–44.

Morris, P. W., & Hough, G. H. (1987). *The anatomy of major projects*. New York: Wiley.

Nguyen, L. D., Ogunlana, S. O., & Lan, D. T. X. (2004). A study on project success factors in large construction projects in Vietnam. *Engineering, Construction and Architectural Management, 11*(6), 404–413.

Pinto, J. K., & Slevin, D. P. (1989). Critical success factors in R&D projects. *Research Technology Management, 32*(1), 31–35.

Pinto, J. K., & Slevin, D. P. (1988a). Project success: Definitions and measurement techniques. *Project Management Journal, 19*(1), 67–72.

Pinto, J. K., & Slevin, D. P. (1988b). Critical success factors across the project life cycle. *Project Management Journal, 19*(3), 67–74.

Rockart, J. F. (1982). The changing role of the information systems executive: A critical success factors perspective. *Sloan Management Review Fall, 24*, 3–13.

Rowlinson, S. (1999). Selection criteria. In S. Rowlinson & P. McDermott (Eds.), *Procurement systems: A guide to best practice* (pp. 276–299). London: E and F.N. Spon.

Rubin, I. M., & Seeling, W. (1967). Experience as a factor in the selection and performance of project managers. *IEEE Transactions on Engineering Management, 14*(3), 131–134.

Sayles, L. R., & Chandler, M. K. (1971). *Managing large systems*. New York: Harper and Row.

Schaufelberger, J. E. (2003) Success factors for design-build contracting. *Conference Proceeding Paper, Construction Research Congress—Wind of Change: Integration and Innovation of Construction ASCE Conference Proceedings*. doi: http://dx.doi.org/10.1061/40671(2003)89.

Schultz, R. L., Slevin, D. P., & Pinto, J. K. (1987). Strategy and tactics in a process model of project implementation. *Interfaces, 17*(3), 34–46.

Thompson, P. (1991). The client's role in project management. *International Journal of Project Management, 9*(2), 90–92.

Chapter 4
Critical Success Factors for Projects

Abstract The twenty success/failure factors derived in Chap. 3 have been evaluated for their criticality in influencing the project outcome in terms of performance on schedule, cost, quality, and no-dispute. For the evaluation, data has been collected through a second stage questionnaire survey. For the analysis, multinomial logistic regression (explained briefly in this chapter) has been used. Knowing the project performance to be at a certain performance level it is now possible to identify the most critical factor to focus on for either sustaining the performance or enhancing the performance to a higher level. The analysis has established that the 3C's (commitment, coordination, and competence) are the key factors for achieving schedule, cost, and quality objectives respectively.The research findings are validated through two fold approach: validation through case study and validation through structured interview. A total of twenty real life cases were studied and the research findings were found to be valid and consistent.

4.1 Introduction

In the previous chapter, the success and failure factors corresponding to the four performance criterion: schedule, cost, quality, and no-dispute, were found out from the responses of first stage questionnaire. Thereafter a pool of 20 factors (11 success and 9 failure factors) was created from the union of success and failure factors for the four performance criteria. In this chapter our objective is to assess the criticality of the success and failure factors. For this a second stage questionnaire was conducted. Details of second stage questionnaire were already presented in Chap. 2.

This chapter is based on (a) Commitment, coordination, competence and the iron triangle, International Journal of Project Management, 25(5), 2007, with permission from Elsevier, and (b) Critical Factors Affecting Quality Performance in Construction Projects, Total Quality Management and Business Excellence, 17(9),2006, with permission from Taylor and Francis.

K. N. Jha, *Determinants of Construction Project Success in India*, 85
Topics in Safety, Risk, Reliability, and Quality 23, DOI: 10.1007/978-94-007-6256-5_4,

In the second stage questionnaire, an 11-point scale was used to measure the extent of contribution of the 20 factors on the performance of the 'choice project' while the performance rating of choice project is obtained on a 10-point scale. The meaning of the 'choice project' has already been explained earlier. In the 11-point scale responses were sought from −5 to +5 with '0' in between. While +5 indicated high position contribution, −5 indicated high negative contribution, and '0' meaning no effect. The intermediated values were to be interpreted suitably between the two extremes of high positive to high negative contribution. In the 10-point scale (1–10), 1 represented 'very poor performance' and '10' represented 'very good performance' and the intermediate values were to be interpreted between these two extremes.

For the multiple regression analysis, it was assumed that the performance rating is the function of the 20 factors as shown in the Eq. (4.1). From the equation, it can be easily observed that while the 'performance rating' is treated as a response or dependent variable, the 20 factors have been considered as explanatory or independent variables.

$$\text{Performance rating} = f\,(F1,\ F2,\ F3,\ \ldots\ldots\ldots,\ F20) \tag{4.1}$$

It was pointed out earlier that the response on 'performance rating' of the 'choice project' has been sought on a 10-point scale, this variable can be regarded as a discrete variable, and thus the application of multinomial logistic regression is considered most appropriate as an analysis tool for this case. When this analysis was performed on keeping all the 20 factors together on the right hand side of the Eq. (4.1), none of the factors has emerged as significant although the percentage prediction using this model is 100 %. As suggested by statisticians that the inclusion of irrelevant variables can result in poor model fit, number of variables should be restricted (Whitehead 1998; SPSS Release 9 manual). This is very much similar to step wise regression in case of linear multiple regression. Thus, instead of all the 20 factors in the right hand side of the Eq. (4.1), only those factors which emerged from the corresponding performance criterion were used in the multinomial logistic regression. In the left hand side of the Eq. (4.1), the performance rating of the project under the corresponding criterion is used. In the following sections, we discuss the criticality of success and failure factors corresponding to the schedule, cost, quality, and no-dispute performance criteria. In order to help the readers in understanding the discussion, an illustration for interpreting the multinomial logistic regression is provided.

4.2 Interpretation of Multinomial Logistic Regression Models

The multinomial logistic regression is an extension of binomial logistic regression and the chances of occurrence of a particular value of response variable are compared with the chances of occurrence of the reference value of the response variable, the performance level of the project. In the present study, since our interest is to look for the best outcome and identify those explanatory variables that would enhance the chances of bringing the performance level nearer to the best outcome as well as to identify those that drag the performance level down, the reference value of the performance level is set as 10 i.e., the "very good performance" rating.

The interpretation of the results of this regression is drawn from mainly three components: "odds ratio", generally written as e^B; "log of odds ratio", B; and the "current value" of the explanatory variable which is being compared with the reference value 10.

Odds ratio: It is the ratio of likelihood of occurrence of an event to the likelihood of nonoccurrence of that event and it is denoted by e^B. In order to understand it better, let's assume that 'M' and 'N' represent the values of response or independent variable. More specifically let's consider, that 'N' represents the 'reference level—10' in this study and 'M' represents occurrence of project performance of some desired level called 'current value' having values as 2, 3 and so on up to 9. Thus, if the chances of occurrence of a current value of '2' is 'p', the chances of nonoccurrence of '2' will be '$q = (1 - p)$'. For the case of binomial logistic regression,the $e^B = p/q = p/(1 - p)$. The values of p and q can be determined from e^B. For example, $p = e^B/(1 + e^B)$ and $q = 1/(1 + e^B)$.

Log of odds ratio: It is denoted by B and as the name suggests it is the log of e^B, odds ratio. This component is regarded more for its sign, which determines the impact of explanatory variable on the outcome of response variable. For the event M (the assumed value of response variable), if the analysis shows positive sign to B, it implies that any increase in the value of explanatory variable will increase the likelihood of event being M. Conversely, the negative value of B indicates that increase in the value of explanatory variable will decrease the likelihood of event being M, i.e., occurrence of the response variable being at the current level. Since the performance level is compared with 10, decrease in the likelihood value of performance rating at the current level will indicate the increase in the likelihood value of reference performance rating and vice versa.

The magnitude of impact of explanatory variable on the current value of the response variable is determined by the magnitude of the odds ratio, e^B. More precisely, one unit increase in the value of explanatory variable causes odds ratio to change by $(1 - e^B)$ times, i.e., the new or changed value of odds ratio would now be $e^B\{1 - (1 - e^B)\} = e^{2B}$. Accordingly, the new value of likelihood of event M, p' (say) and that of event N, q' (say) after change due to one unit of explanatory variable will be $e^{2B}/(1 + e^{2B})$ and $1/(1 + e^{2B})$ respectively. If Δp and Δq be the changes in the values of likelihood of events M and N, they can be written as given below.

$$\Delta p = p_{new} - p_{old} = p' - p = \frac{e^{2B}}{1 + e^{2B}} - \frac{e^B}{1 + e^B} \tag{4.2}$$

$$\Delta q = q_{new} - q_{old} = q' - q = \frac{1}{1 + e^{2B}} - \frac{1}{1 + e^{B}} \tag{4.3}$$

As discussed above in this section, the M and N in the present study represent the occurrence of desired level of project performance and N, the reference level of 10. Δp would indicate the change in likelihood of project performance being at the current level and Δq would indicate the change in likelihood of project performance of not being at the current level, i.e., being at the reference level of 10. The values of Δp and Δq are thus complementary to each other. It could be further interpreted that the negative value of Δp which indicates decreasing chances of the project performance being at the current level, is also associated with the positive value of Δq indicating increasing chances of alternate event, i.e., performance level being at 10. These lead to conclude that negative value of Δp indicates improvement in the performance level toward 10 from the current level. On the other hand, positive value of Δp indicates increasing chances of performance of the project being at the same level and decreasing chances of performance being at the alternate level of 10. These lead to conclude that with positive value of Δp there will be diminishing chances of further improvement. These logics are used for interpretations of results of statistical analyses of responses as discussed below.

4.3 Schedule Criterion

4.3.1 Analysis of Responses to Success Factors

It may be recalled that seven success factors (F_1–F_6 and F_9) emerged from the factor analysis conducted on success attributes. For the application of multinomial logistic regression, these success factors have been considered as the explanatory variables and the responses to actual schedule performance of the choice project have been taken as the response variables.

Out of the seven factors, only three factors namely-F5-Commitment of project participants, F_6-owner's competence, and F_9-Good coordination among project participants) have emerged to be significant at different schedule performance rating as shown in Table 4.1. In the analysis the Chi square value of 143.986 with 63 degree of freedom is highly significant (significance level 0.000). According to Whitehead (1998), this is an indication that the null hypothesis that all effects of the independent variables are zero can be rejected.

The three R^2 measures available in the SPSS software are: Cox and Snell R^2, Nagelkerke R^2, and McFadden R^2. The R^2 measures confound goodness of fit and explanatory power of the model. However, statisticians recommend the Nagelkerke R^2 value as the most relevant value to report. In this case it is found to be 0.827, which indicates that the model performs well for the schedule performance rating ranges of 5, 8 and 9 with rating 10 being the reference category. The percent correct prediction is 51.7 %.

Table 4.1 Summary of important results of multinomial logistic regression between success factors and schedule performance rating (Iyer and Jha 2006, with permission from ASCE)

Schedule performance rating (SPR) (1)	Variable (2)	Log of odds ratio, B (3)	Std. error, SE (4)	Wald stat = $(B/SE)^2$ (5)	Sig. level, α (6)	Odds ratio, e^B (7)	p (8)	q (9)	Δp (10)	Δq (11)
5.00	Intercept	4.226	4.983	0.719	0.396					
	F_5	−1.749	0.803	4.739	0.029	0.174	0.15	0.85	−0.12	0.12
6.00	Intercept	10.140	3.938	6.631	0.010					
7.00	Intercept	8.119	2.714	8.947	0.003					
8.00	Intercept	6.810	2.601	6.856	0.009					
	F_5	−1.149	0.486	5.596	0.018	0.317	0.24	0.76	−0.15	0.15
9.00	Intercept	−3.115	4.237	0.541	0.462					
	F_6	−1.055	0.425	6.169	0.013	0.348	0.26	0.74	−0.15	0.15
	F_9	1.689	1.028	2.703	0.100	5.416	0.84	0.16	0.12	−0.12

In the present case the negative value of B corresponding to performance rating of 5 in Col 3 of Table 4.1, indicates that any increase in variable value of F_5 would significantly decrease the likelihood of project schedule performance rating being at the level 5. The values of Δp and Δq in Col 10 and Col 11 indicate that with one unit rise in the value of F_5 the probability of performance to remain at level 5 decreases by 12 % or conversely the probability of achieving the performance level 10 (bettering the performance) will increase by 12 %.

It can also be seen from Col 7 that values of odds ratio of F_5 neither remain same nor significant in other schedule performance ratings, e.g., while it has the value of 0.317 at the performance rating of 8, it does not appear as significant variable in other performance ratings. It indicates that at a current performance rating level of 8, one unit increase in the value of F_5 will enhance the probability of bettering the performance by 15 % (Col 10 and Col 11 in Table 4.1).

Similarly, at the performance rating level of 9, two variables, F_6 (owner's competence) and F_9 (coordination among project participants) are observed to be significant with their odd ratios as 0.348 and 5.416 respectively. The B-value of F_6 is −1.055 which indicate that it is only owner's competence that can play vital role in further betterment of project performance with probability of enhancement being 15 % (Δq value). All these results, besides supporting the conclusions of previous paragraph on commitment of project participants, also lead to conclude that when the schedule performance rating is at very high level it is the owner who should take lead in guiding the project team as their role has high impact on taking the project at still higher level of schedule performance. At the very high performance level, obviously all factors by itself must be working at its best, but probably the owner's competence and guidance will trigger for further betterment.

Similarly, the B-value of F_9 being 1.689 at the performance rating level 9 indicates that an increase in the factor value of 'coordination among project participants' by one unit will retain the project at performance level of 9 with an increased probability (Δp = 12 %) and diminishing chances of further improvement to 10. In other words, this could also be interpreted that decrease in coordination among participants by one unit may lead the performance rating sliding down from the current level of 9. Any project manager would be quite cautious to see the level of performance does not at least slide down, if not improving. In the particular case the performance being retained at 9 is a quite a satisfactory and healthy symbol and effort should be put to further improve or at least retain the current level of coordination to maintain the current performance level.

4.3.2 Analysis of Responses to Failure Factors

Similar analysis is carried out taking seven failure factors (F_{12}–F_{18}) as explanatory variables and schedule performance ratings as response variable. It is found that four variables (F_{12}-Conflict among project participants, F_{13}-PM's ignorance and lack of knowledge, F_{14}-Hostile social environment, and F_{17}-Harsh climatic condition at site) are significant variables among the seven failure variables at different schedule performance ratings. The important results of the analysis are summarized in Table 4.2. In the analysis the Chi square value of 108.099 with 63 degree of freedom is highly significant (significance level of 0.000). The Nagelkerke R^2 value is 0.723 and percent correct prediction is 47 %.

Similar to previous discussion the negative value of B of F_{12} corresponding to performance rating of 3 (in Col 3 of Table 4.2) indicates that any increase in variable value of F_{12} (Conflict among project participants) would significantly decrease the likelihood of project schedule performance remaining at level 3, but increase the chances of bettering. The values of Δp and Δq in Col 10 and Col 11indicate that one unit rise in the value of F_{12}, likelihood of bettering the performance will be 10 % when the present performance level is 3. This result that chances of bettering the performance with conflict among project participants however appears contradictory to the common belief. But some researchers in the past have also concluded that conflict among project participants can yield positive results. Tjosvold (1991), Deustch (1994), Crowley and Karim (1995) and Cheung and Chuah (1999) suggest that when conflict is appropriately managed, it could be constructive and even add substantial value to an organization. It leads to creative solutions, which enhance the ability to work together in the future. Kumaraswamy (1998) has also concluded that while conflict is inevitable on construction projects, it is necessary for management to differentiate destructive from constructive conflict, and to anticipate and minimize the former, while carefully controlling the latter.

Similarly positive B-value of F_{13} (project manager's ignorance and lack of knowledge), F_{14} (hostile socio-economic environment) and F_{16} (indecisiveness of project participants) indicate that unit increase in their values will increase the likelihood of project schedule performance remaining at the lower level compared to the reference level of 10. Probability values of worsening can be gaged from the Δp values given in Col 10 of Table 4.2. Based on these findings, the project

Table 4.2 Summary of important results of multinomial logistic regression between failure factors and schedule performance rating (Iyer and Jha 2006, with permission from ASCE)

Schedule performance rating (SPR) (1)	Variable (2)	Log of odds ratio, B (3)	Std. error, SE (4)	Wald stat = $(B/SE)^2$ (5)	Sig. level, α (6)	Odds ratio, e^B (7)	p (8)	q (9)	Δp (10)	Δp (11)
3.00	Intercept	−2.161	1.123	3.704	0.054					
	F_{12}	−1.958	1.051	3.472	0.062	0.141	0.12	0.88	−0.10	0.10
	F_{13}	1.503	0.769	3.820	0.051	4.496	0.82	0.18	0.13	−0.13
4.00	Intercept	−9.123	5.461	2.791	0.095					
	F_{14}	1.164	0.560	4.322	0.038	3.202	0.76	0.24	0.15	−0.15
	F_{16}	−2.346	1.408	2.776	0.096	0.096	0.09	0.91	−0.08	0.08
5.00	Intercept	−5.921	4.327	1.872	0.171					
	F_{16}	3.212	1.803	3.172	0.075	24.824	0.96	0.04	0.04	−0.04
6.00	Intercept	−1.977	1.016	3.783	0.052					
	F_{12}	−2.166	0.924	5.487	0.019	0.115	0.10	0.90	−0.09	0.09
	F_{13}	1.162	0.645	3.247	0.072	3.195	0.76	0.24	0.15	−0.15
	F_{16}	1.129	0.614	3.385	0.066	3.092	0.76	0.24	0.15	−0.15
9.00	Intercept	−0.646	0.565	1.310	0.252					
	F_{13}	−0.673	0.360	3.501	0.061	0.510	0.34	0.66	−0.13	0.13

professionals can be educated to be more careful in handling these variables lest the project performance is likely to slip down.

4.4 Cost Criterion

4.4.1 Analysis of Responses to Success Factors

Analysis of responses on success factors (F_1–F_7 and F_9: eight factors) on the performance of cost criterion of the choice project identifies four factors (F_2, F_4, F_5 and F_9) to be significant. Important results are summarized in Table 4.3. In the analysis the Chi square value of 68.163 with 49 degree of freedom is highly significant (significance level of 0.030). The Nagelkerke R^2 value is 0.571 and percent correct prediction is 42 %.

In the present case the negative value of B corresponding to performance rating of 4 in Col 3 of the Table indicates that any increase in variable value of F_9 "Good coordination among project participants" would significantly decrease the likelihood of project cost performance rating remaining at current level 4, but increase the probability of bettering. The values of Δp and Δq in Col 10 and Col 11 indicate that with unit rise in the value of F_9, likelihood of bettering the performance will be 9 % when the present performance level is 4.

Similarly, the negative value of B corresponding to performance rating of 5 and 6 in Col 3 of Table 4.3 for F_4 (Favorable working condition) also indicate that any

Table 4.3 Summary of important results of multinomial logistic regression between success factors and cost performance rating

Cost performance rating (CPR) (1)	Variable (2)	Log of odds ratio, B (3)	Std error, SE (4)	Wald stat = $(B/SE)^2$ (5)	Sig. level, α (6)	Odds ratio, e^B (7)	p (8)	q (9)	Δp (10)	Δq (11)
4.00	Intercept	2.891	4.806	0.362	0.548					
	F_9	−2.168	1.205	3.236	0.072	0.114	0.10	0.90	−0.09	0.09
5.00	Intercept	2.567	4.775	0.289	0.591					
	F_2	1.903	1.098	3.006	0.083	6.708	0.87	0.13	0.11	−0.11
	F_4	−0.908	0.492	3.407	0.065	0.403	0.29	0.71	−0.15	0.15
6.00	Intercept	5.870	3.475	2.854	0.091					
	F_4	−0.809	0.477	2.874	0.090	0.445	0.31	0.69	−0.14	0.14
	F_5	1.303	0.758	2.957	0.086	3.680	0.79	0.21	0.14	−0.14
	F_9	−1.442	0.859	2.819	0.093	0.236	0.19	0.81	−0.14	0.14
7.00	Intercept	5.152	3.131	2.709	0.100					
8.00	Intercept	5.199	2.811	3.421	0.064					
	F_2	0.845	0.496	2.904	0.088	2.329	0.70	0.30	0.14	−0.14
9.00	Intercept	5.700	2.827	4.065	0.044					
	F_5	0.930	0.466	3.975	0.046	2.534	0.72	0.28	0.15	−0.15

increase in variable value of F_4 would significantly increase the likelihood of bettering the project cost performance rating from the current level by 14–15 %.

Further it is also observed from Table 4.3 that no variable has been found to be significant at performance rating of 7, but at performance rating level of 8 and 9, two variables, F_2 (Top management support) and F_5 (Commitment of all project participants) are observed to be significant with their odd ratios as 2.329 and 2.534 respectively. Their corresponding B-values of 0.845 and 0.930 indicate that increase in F_2 and F_5 will retain the project at current performance level with increased probabilities by 14 and 15 % respectively. Since the current performance level (8 or 9) is also quite high, the variables can be considered to be very important and necessary efforts should be put to improve or at least retain the top management support and commitment of project participants to have high level of project performance.

4.4.2 Analysis of Responses to Failure Factors

Similar analysis is carried out taking seven failure factors (F_{12}–F_{14}, F_{16} to F_{18}, and F_{20}) as explanatory variables and cost performance ratings as response variable. It is found that four variables (F_{12}-Conflict among project participants, F_{13}-PM's ignorance and lack of knowledge, F_{16}-Indecisiveness of project participants, F_{17}-Harsh climatic condition at site, and F_{18}-Project specific factor) are significant variables among the seven failure variables at different cost performance ratings. The important results of the analysis are summarized in Table 4.4. In the analysis

Table 4.4 Summary of important results of multinomial logistic regression between failure factors and cost performance rating

Cost performance rating (CPR) (1)	Variable (2)	Log of odds ratio, B (3)	Std error, SE (4)	Wald stat = $(B/SE)^2$ (5)	Sig. level, α (6)	Odds ratio, e^B (7)	p (8)	q (9)	Δp (10)	Δq (11)
3.00	Intercept	−129.214	0.000							
4.00	Intercept	−2.155	1.289	2.792	0.095					
	F_{12}	−1.427	0.796	3.215	0.073	0.240	0.19	0.81	−0.14	0.14
	F_{13}	1.889	0.802	5.551	0.018	6.610	0.87	0.13	0.11	−0.11
5.00	Intercept	−1.587	1.120	2.008	0.156					
	F_{18}	−0.672	0.322	4.363	0.037	0.511	0.34	0.66	−0.13	0.13
6.00	Intercept	−2.102	1.359	2.394	0.122					
	F_{12}	−2.035	1.060	3.684	0.055	0.131	0.12	0.88	−0.10	0.10
	F_{13}	1.358	0.725	3.504	0.061	3.888	0.80	0.20	0.14	−0.14
	F_{17}	0.942	0.521	3.273	0.070	2.564	0.72	0.28	0.15	−0.15
7.00	Intercept	−0.410	0.664	0.380	0.537					
	F_{16}	0.800	0.465	2.965	0.085	2.226	0.69	0.31	0.14	−0.14
8.00	Intercept	1.067	0.493	4.687	0.030					
	F_{13}	0.461	0.278	2.747	0.097	1.585	0.61	0.39	0.10	−0.10
	F_{17}	−0.527	0.298	3.131	0.077	0.590	0.37	0.63	−0.11	0.11
	F_{18}	−0.506	0.188	7.231	0.007	0.603	0.38	0.62	−0.11	0.11

the Chi square value of 76.056 with 49 degree of freedom is highly significant (significance level of 0.008). The Nagelkerke R^2 value is 0.614.

It can be observed from Col 3 of Table 4.4 that the log of odds ratio of F_{12} has been negative which indicates that with increase in value of F_{12} there are enhanced chances of bettering the cost performance of the project from the current performance level of 4 or 6. The probability of enhancement will however be 14 and 10 % respectively from the levels 4 and 6. This factor has also been observed to be significant in the schedule performance criterion at performance levels of 3 and 6 (Table 4.2). Further this factor does not seem to be significant at higher performance levels. All these lead to conclude that the importance level of this factor can be more realized and appreciated at lower performance level than at higher performance level.

The other variable F_{13} (PM's ignorance and lack of knowledge) is observed to be significant (Table 4.4) at performance rating levels of 4, 6 and 8. The positive value of Δp indicates that unit increase in this variable would only increase the probability of the performance rating being at the same level rather than improving. As overall objective of the project will always be to better the performance, this factor F_{13} should be given due importance as existence of this factor would not allow the project performance to improve and thus a competent project manager should be posted to handle the project.

Similarly inferences can be drawn from the positive values of Δp of variables F_{16}, F_{17} and F_{18} that indecisiveness of project participants (F_{16}), and harsh climatic condition at site (F_{17}) and Project specific factor (F_{18}) would not let project performance improve from its existing levels. However, the negative value of Δp of F_{17} and F_{18} at project performance rating 8 remains unexplainable.

4.5 Quality Criterion

4.5.1 Analysis of Responses to Success Factors

The multinomial logistic regression between the quality performance rating as the response variable and the success factors corresponding to quality performance in factor analysis viz. F_1, F_2, F_3, F_6, F_7, and F_8 as explanatory variables is carried out. A summary of analysis results are produced in Table 4.5. In the analysis the Chi square value of 65.605 with 36 degree of freedom is highly significant (significance level of 0.002). The Nagelkerke R^2 value is 0.556 and percent correct prediction is 44 %.

It is found from the analysis that none of the variables appear significant when the project quality is at lower level i.e. below level 5. Only four factors F_1, F_2, F_7 and F_8 are found to be significant at different performance ratings. Even among these factors, F_1 (project manager's competence) has been found to have emerged in four performance levels, 5, 6, 8 and 9. Its Δp and Δq values (Col 10 and Col 11)

Table 4.5 Summary of important results of multinomial logistic regression between success factors and Quality performance rating (Jha and Iyer 2006, with permission from Taylor and Francis)

Quality performance rating (QPR) (1)	Variable (2)	Log of odds ratio, B (3)	Std. error, SE (4)	Wald stat = $(B/SE)^2$ (5)	Sig. level, α (6)	Odds ratio, e^B (7)	p (8)	q (9)	Δp (10)	Δq (11)
5.00	Intercept	7.077	4.428	2.555	0.110					
	F_1	−1.780	1.005	3.138	0.076	0.169	0.14	0.86	−0.12	0.12
6.00	Intercept	4.931	3.974	1.540	0.215					
	F_1	−2.406	0.935	6.617	0.010	0.090	0.08	0.92	−0.07	0.07
	F_6	3.383	1.774	3.638	0.056	29.459	0.97	0.03	0.03	−0.03
7.00	Intercept	3.794	2.483	2.335	0.126					
	F_2	−1.000	0.543	3.386	0.066	.368	0.27	0.73	−0.15	0.15
	F_7	1.534	0.682	5.055	0.025	4.635	0.82	0.18	0.13	−0.13
8.00	Intercept	3.503	1.845	3.604	0.058					
	F_1	−0.926	0.419	4.879	0.027	0.396	0.28	0.72	−0.15	0.15
	F_8	−0.613	0.318	3.719	0.054	0.542	0.35	0.65	−0.12	0.12
9.00	Intercept	0.865	1.908	0.205	0.650					
	F_1	−0.852	0.405	4.439	0.035	0.426	0.30	0.70	−0.15	0.15

indicate that one unit rise in the value of F_1, likelihood of bettering the performance will be 12, 7, 15, and 15 % respectively when the present performance level is 5, 6, 8, and 9 respectively. Even if the percentage increase in the likelihood of bettering is taken as indicative, it still leads to conclude that the factor F_1 is the most important of all factors to achieve improved performance. Highest level of importance to project manager's competence, as obtained in this study, for achieving quality also tends to reestablish the philosophy of quality gurus like Deming, Crosby, Juran and Taguchi that key to quality lies with management than with workforce.

It can be also observed from Table 4.5 that by increasing top management support (F_2), likelihood of bettering the performance will be 15 % from its current level of 7. Also when the project quality is at level 8 a unit increase in the interaction between project participants (F_8) would result in 12 % increase in the probability of producing the 'very good' (Level 10) quality. The meanings could not be assigned to the positive B value of variable F_6 and F_7 in level 6 and level 7 as they indicate that increase in owner's competence and internal interaction between project participants are more likely to keep the project quality at the same level.

4.5.2 Analysis of Responses to Failure Factors

The application of multinomial regression for failure factors corresponding to quality performance viz. F_{12}, F_{13}, F_{14}, F_{17}, F_{18}, and F_{20} does not result into a significant model. The Chi square value of 38.782 with 36 degree of freedom is significant at 0.345, which means that the null hypothesis that all effects of the independent variable are zero can be accepted. Hence the model formed in this case is not fit to be interpreted and hence discarded.

4.6 No-Dispute Criterion

4.6.1 Analysis of Responses to Success Factors

The success factors corresponding to this performance criterion are F_1, F_2, F_4, F_6, F_{10}, and F_{11}. Application of multinomial logistic regression suggests that only three factors (F_2, F_4 and F_6) out of the six contribute significantly in preventing the disputes. Table 4.6 shows the summary of results of multinomial logistic regression between success factors and no-dispute performance ratings. As can be seen the Chi square value of 72.607 at 54 degree of freedom is significant at 0.046. The Nagelkerke R^2 value is 0.587 and percent correct prediction is 34 %. It can also be observed that no variable has emerged to be significant at lower and higher no-dispute performance ratings.

In the present case the negative values of B corresponding to performance rating of 5 (Col 3 of Table 4.6), indicates that any increase in variable value of F_2 (Top management support), F_4 (Favorable working condition), and F_6 (Owner's competence) would significantly decrease the likelihood of project disputes remaining at the same level. The odds ratios of 0.131, 0.208, and 0.217 in Col 7 coupled with Δp or Δq values in Col 10 and Col 11, indicate that one unit rise in the value of F_2, F_4, and F_6 will result in 10, 13, and 13 % increase in the probability of bettering the no-dispute performance respectively when the present performance level is 5.

Table 4.6 Summary of important results of multinomial logistic regression between success factors and Dispute performance rating

Dispute performance rating (DPR) (1)	Variable (2)	Log of odds ratio, B (3)	Std. error, SE (4)	Wald stat = $(B/SE)^2$ (5)	Sig. level, α (6)	Odds ratio, e^B (7)	p (8)	q (9)	Δp (10)	Δq (11)
4.00	Intercept	6.462	3.843	2.828	0.093					
5.00	Intercept	−0.390	5.821	0.004	0.947					
	F_2	−2.035	1.052	3.738	0.053	0.131	0.12	0.88	−0.10	0.10
	F_4	−1.570	0.818	3.680	0.055	0.208	0.17	0.83	−0.13	0.13
	F_6	−1.529	0.815	3.521	0.061	0.217	0.18	0.82	−0.13	0.13
7.00	Intercept	6.404	2.618	5.985	0.014					
8.00	Intercept	5.652	2.611	4.687	0.030					

4.6.2 Analysis of Responses to Failure Factors

Similar analysis results with failure factors F_{12}, F_{13}, F_{14}, F_{16}, F_{17}, and F_{20} corresponding to no-dispute performance are given in Table 4.7. The Chi square value of 80.434 at 54 degree of freedom is significant at 0.011.The Nagelkerke R^2 value is 0.621 and percent correct prediction is 47 %.

It can be seen that only three factors F_{13}, F_{16}, and F_{17} are found significant out of the six failure factors. While the positive value of Δp of F_{16} can be interpreted that indecisiveness of project participants (F_{16}) may not allow the project performance to improve in the no-dispute criterion, negative values F_{17} and F_{18} cannot be explained.

To recapitulate the preceding discussion, Table 4.8 is provided. This table summarizes the effect of the critical success and failure factors across the four performance criteria at different performance levels.

It can be observed from Table 4.8 that six factors (F_3—monitoring and feedback; F_{10}—availability of trained resources; F_{11}—Regular budget update; F_{15}—owner's incompetence; F_{19}—negative attitude of project participants; and F_{20}—faulty project conceptualization) have not appeared significant in influencing project outcome. However, the importance of these factors has been emphasized and their impacts on project outcomes have been dealt elaborately in the literature (Chan et al. 2001a; Cho and Gibson 2001; Lim and Ling 2002). Reason for differences in findings of this study when compared to other study could be traced as this study has been more generic taking all factors into account, while other studies have been case specific.

Incidentally, none of the factors has been found to have significant influence on all four performance criteria. However, among 11 success factors two factors: F_2—top management support and F_6—owner's competence have been found to be influencing significantly at least three performance criteria. While F_2 contribute in improvement in cost, quality and no-dispute performance criteria, F_6 contribute in schedule, quality and no-dispute performances. Similarly, among the failure factors F_{13}—project manager's ignorance and lack of knowledge; and F_{16}—

Table 4.7 Summary of important results of multinomial logistic regression between failure factors and dispute performance rating

DPR (1)	Variable (2)	Log of odds ratio, B (3)	Std. error, SE (4)	Wald Stat = $(B/SE)^2$ (5)	Sig. level, α (6)	Odds ratio, e^B (7)	p (8)	q (9)	Δp (10)	Δq (11)
6.00	Intercept	−1.970	1.090	3.265	0.071					
	F_{16}	1.540	0.594	6.723	0.010	4.667	0.82	0.18	0.13	−0.13
	F_{17}	−1.503	0.584	6.619	0.010	0.222	0.18	0.82	−0.13	0.13
8.00	Intercept	0.056	0.543	0.011	0.918					
	F_{13}	−0.521	0.284	3.353	0.067	0.594	0.37	0.63	−0.11	0.11
9.00	Intercept	0.798	0.460	3.013	0.083					

Table 4.8 Summary of significant impacts of project success and failure factors on various levels of performance under different performance criteria

Performance criteria	Significantly influenced performance ratings and subsequent impact of the factors											
	Schedule			Cost			Quality			No dispute		
Factors	SPR[a]	Δp^b	Δq^b	CPR[a]	Δp^b	Δq^b	QPR[a]	Δp^b	Δq^b	DPR[a]	Δp^b	Δq^b
F_1							5	−0.12	0.12			
							6	−0.07	0.07			
							8	−0.15	0.15			
							9	−0.15	0.15			
F_2				5	0.11	−0.11	7	−0.15	0.15	5	−0.10	0.10
				8	0.14	−0.14						
F_3	Does not influence project performance in any rating											
F_4				5	−0.15	0.15				5	−0.13	0.13
				6	−0.14	0.14						
F_5	5	−0.12	0.12	6	0.14	−0.14						
	8	−0.15	0.15	9	0.15	−0.15						
F_6	9	−0.15	0.15				6	0.03	−0.03	5	−0.13	0.13
F_7							7	0.13	−0.13			
F_8							8	−0.12	0.12			
F_9	9	0.12	−0.12	4	−0.09	0.09						
				6	−0.14	0.14						
F_{10}	Does not influence project performance in any rating											
F_{11}	Does not influence project performance in any rating											
F_{12}	3	−0.10	0.10	4	−0.14	0.14						
	6	−0.09	0.09	6	−0.10	0.10						
F_{13}	3	0.13	−0.13	4	0.11	−0.11				8	−0.11	0.11
	6	0.15	−0.15	6	0.14	−0.14						
	9	−0.13	0.13	8	0.10	−0.10						
F_{14}	4	0.15	−0.15									

(continued)

Table 4.8 (continued)

Performance criteria	Significantly influenced performance ratings and subsequent impact of the factors											
	Schedule			Cost			Quality			No dispute		
Factors	SPR[a]	Δp[b]	Δq[b]	CPR[a]	Δp[b]	Δq[b]	QPR[a]	Δp[b]	Δq[b]	DPR[a]	Δp[b]	Δq[b]
F_{15}	Does not influence project performance in any rating											
F_{16}	4	−0.08	0.08	7	0.14	−0.14				6	0.13	−0.13
	5	0.04	−0.04									
	6	0.15	−0.15									
F_{17}				6	0.15	−0.15				6	−0.13	0.13
				8	−0.11	0.11						
F_{18}				5	−0.13	0.13						
				8	−0.11	0.11						
F_{19}	Does not influence project performance in any rating											
F_{20}	Does not influence project performance in any rating											

[a] SPR, CPR, QPR and DPR indicate the performance rating levels under schedule, cost, quality and no-dispute criteria

[b] Δp and Δq indicate the incremental changes in probability values of current performance rating being at the same level or improving from the current level

indecisiveness of project participants have been found to be significantly influencing three project performance criteria: schedule, cost, and no-dispute.

4.7 Validation of Results

In Chap. 3 and the previous sections of this chapter, analyses of questionnaire responses were discussed and the inferences were drawn from the results of various analyses. Generally, respondents of the 1st and 2nd stage questionnaires occupied top and responsible positions in their respective organizations and they are all vastly experienced in the construction management field. Also there has been consistency in their responses and the feedback given by the respondents could be assumed to be dependable and their views noteworthy. In similar studies in the past when researchers have collected responses from such responsible respondents, have expressed their contentment and tried to prove that this in itself lends credibility and carries sufficient validation to the research findings (Chan and Kumaraswamy 1997; Ling 2002). Yet separate validation of analysis results has been undertaken in this study.

Case studies to validate the analysis results are one of the most acceptable method of validation (Fellows and Liu 1997; Yin 1984; Jefferies et al. 2002); In order to carry out the case studies, live projects were selected randomly through contacts with construction organizations. Since the study was pertaining to only construction phase of the project, the case projects selected were those, which have already been completed in the recent past or were in advanced stage of completion. The methodology consisted of referring the contract document, quality assurance manual, and the correspondence files of the contracts. Brief backgrounds of the projects are also presented in this chapter.

The case study was followed by the structured interviews with professionals involved in developing the design, budget cost, project plans and construction. Structured interviews are considered as another accepted tool for validation (Songer and Molenar 1997) and this method also provides clarity to the survey results. Structured interviews have been conducted among thirty experienced professionals who are the leading experts in the construction management field. Structured interviews lasted for about half an hour and it sought the views of interviewees on a range of issues. The extract of the relevant portions of the structured interview is shown in the Box 4.1. Each of the interviewees was given research brief in which they were introduced to success and failure factors. They were told to rank the set of success factors derived from this research in order of the importance. During the course of the interviews, interviewees were not told about the findings of the research as it was felt that revelation of the research results might create a bias in the interviewee's responses.

Box 4.1 Format of Structured Interviews

The following success factors have been derived from the research conducted by us based on the analyses of data collected through two stages of questionnaire survey. If you were to choose from among the following factors what would be your order of preference to ensure overall project success. Please rank the following in order of the importance.

S. No.	Description of success factors	Your order of preference
1	Project manager's competence	
2	Top management support	
3	Monitoring and feedback by project participants	
4	Favorable working condition	
5	Commitment of all project participants	
6	Owners competence	
7	Interaction between project participants-internal	
8	Interaction between project participants-external	
9	Availability of trained resources	
10	Regular budget update	
11	Good coordination between project participants	

If schedule achievement is your main priority what 3 factors will you choose from among the above factors in order of preference?

1. ——— 2.——— 3.———

If executing project within cost is your main priority what 3 factors will you choose from among the above factors in order of preference?

1. ——— 2.——— 3.———

If compliance to quality is your main priority what 3 factors will you choose from among the above factors in order of preference?

——— 2.——— 3.———

If you were to avoid dispute between other project participants such as client/contractor/and consultant, what 3 factors will you choose from among the above factors in order of preference?

——— 2.——— 3.———

The broad groups of activities that are performed by the project coordinator in order to achieve the day-to-day coordination are given here. Please rank the following in order of the importance. Rank one indicates most important factor, two next important and so on.

S. No.	Broad group of activities	Rank
1	Resource handling	
2	Team building	
3	Contract Implementation	
4	Planning	

4.8 Conclusions From Case Studies

A total of 20 cases have been studied and due to difficulty in getting access to data, most cases are taken from one single contracting company, which has kindly let the researcher to access their files. This company is part of $ 2bn India's largest engineering and construction conglomerate. It offers turnkey construction services and engineered turnkey industrial and infrastructure projects in civil, mechanical, electrical and instrumentation engineering. The other companies gave only a limited access, but encouraged general discussions that led to the idea of structured interview. The case projects are referred to as P1, P2, ..., P20 and their names are given below.

P1 Mega Housing Project of Delhi Development Authority at Dwarka
P2 Construction of Compressors Factory at Ballabhgarh, Haryana of M/s Techumseh India Pvt Ltd
P3 Construction of Santosh Medical College at Ghaziabad of Maharaja Education Trust
P4 Construction of OrientalCollege at Bhopal
P5 Construction of Grasim Cement Factory at Chittorgarh of Birla Group
P6 Construction of Natural Draft Cooling Towers at Rawatbhata for Nuclear Power Corporation
P7 IMCC Project at New Delhi for Delhi Metro Rail Corporation
P8 Construction of Indira Gandhi International Airport Terminal Building and Extension of Flyover at New Delhi for Airport Authority of India Limited
P9 Construction of New Nizamuddin Bridge at New Delhi for Ministry of Surface Transport, Govt of India
P10 Construction of Hitech City at Hyderabad for L&T-APIIC
P11 Hockey Stadium at Hyderabad for Sports Authority of India Limited
P12 Construction of Nehru Stadium at Chennai
P13 Construction of Assembly Buildings at Gurgaon for Maruti Udyog Limited
P14 Construction of Moser Baer Factory Building at Greater NOIDA
P15 Construction of Detergent Factory for Proctor & Gamble
P16 Construction of Parliament Library Building for Parliament of India
P17 Construction of Tarapore Atomic Power Plant for Nuclear Power Corporation
P18 Construction of Ramagundam Super Thermal Power Project
P19 Construction of Sri Satya Sai Telgu Ganga Project
P20 Construction of Hero Cycles Factory at Ludhyana

The brief background of 12 of them is presented in the following paragraphs for greater clarity.

Box 4.2 Details of Project P1-Mega Housing Project

Delhi Development Authority (DDA) commenced its housing activities in 1967 and since then has played a crucial role in providing more than a million houses to the people of Delhi. They have been constructing houses in Delhi according to the requirements and purchasing capacity of different strata of society. In order to meet the growing demand and to clear the housing backlog DDA has started construction of MIG (middle income group) and LIG (low income group) houses on a large scale. One such project known as Mega Housing Project is currently being executed. The project includes the construction of 504 MIG and 360 LIG Flats along with necessary infrastructure at Dwarka a suburban area of Delhi. This project is now completed.

The project involves 40,000 m^3 of Reinforced Concrete (RC), 3,300 t of rebar, 370,000 m^2 of Formwork, and 365 t of Structural Steel work. The project was awarded to the lowest bidder under competitive bidding. The contract value for the project is Rs. 394 million and the contract duration is 34 months. During the case study it was found that the contract period was over and the project was not complete then. The contractor was given extension of time for 8 months. In this project the client insisted the use of aluminum formwork for construction, which is not locally available and has to be imported from some other countries. This type of formwork was being used by both client and contractor for the first time and none had the background or know-how of the formwork Only 2,200 m^2 formwork was projected by the by the head quarter engineers as against actual estimation of 3,500 m^2 by the site. Head quarter engineer did not contact site office and the procurement action was taken for a far less quantity against the actual requirement. Besides, as per the contractor organization's internal policy procurement of this system formwork had to be routed through the contractor's head quarter. The procurement item being new, design teams of head quarter and regional offices were also involved in addition to the involvement of routine procurement staff at head quarter, but no coordinator was nominated for interaction among the people involved in procurement. With the involvement of so many people and without a central coordinator; there was a lack of communication between various levels of project staff such as site office not contacted for the actual requirement causing inadequate quantity of formwork being ordered; or site office being unaware of delivery schedule of the formwork at site.

It can be observed from the description of P1 (Box 4.2), the error in estimation of formwork quantity due to wrong assumption of cycle time, and the delay in formwork material procurement resulted in delay in completion of the project. Besides the technology was new for the contractor, which he was unaware of. Since the contractor did not even know the manufacturer and supplier of such formwork the procurement action got delayed. After placing the order with one of the international suppliers there was further delay in delivery due to poor coordination between contractor and the supplier. The formwork system being newly implemented by the contractor, it took considerable time to get accustomed by the supervising staff and workmen. These causes can be attributed to lack of coordination on part of contractor and supplier; and inexperience of contractor for such special job. In addition to this, there was an initial delay on part of owner in issuing construction drawings. The owner representative also showed little commitment as they had the habit of postponing the important decisions such as pouring of concrete after formwork and rebar were ready. All these delays also led to increase in cost. Some other reasons for increase in cost were the high increase in material cost, which could not be compensated even though the escalation clause exists in the contract condition. Thus the contractor suffered loss and the project had both time and cost overruns. This project is therefore considered as a failure project and the reasons for the failure could be traced as *lack of coordination on part of contractor and supplier; inexperience of contractor for such special job;* and *lack of commitment* on part of owner.

Box 4.3 Project P2-Compressors Project Ballabhgarh, Haryana

Tecumseh India Private Limited intended to put up a compressor factory at 38 km stone, Mathura road, Ballabhgarh. They appointed C R Narayana Rao as the architect for this project. The architect on behalf of the owner invited the bid from five leading contractors of the country in the month of Dec 1997 and the project was awarded to Larsen and Toubro Limited on 14 Jan 1998 at an amount of Rs. 76.4 million after two rounds of negotiation. It was a time bound project and contractor was given milestone dates for the completion of different structures. The project consisted of construction of civil work, structural, roads, overhead tanks, underground sump, tube wells and architectural works. As per the original schedule of completion the entire project was to be completed by 1 Sep 1998 except the two major buildings named Engineered Buildings 1 and 2, which were to be completed by 30 June 1998. Although the contractor had mobilized the site within the stipulated duration, the work could not be started until the second week of March 1998 resulting into a delay of about 45 days in the beginning of the project itself. The project was finally handed over to the owners in Feb 1999. Along with the final bill the contractor raised claims worth Rs 60 million under three broad categories: Idling charges, escalation, and extended stay citing the three reasons: (1)Initial delay of about 45 days in giving clearance for start of work, (2) Non-release of structures/activities, and (3) Drastic reduction in the scope of works for some items

It was observed from various correspondences of the contractor with the owner that these claims were mainly focused on account of late front clearance and hold in important decision as well as delayed decisions. For some of the buildings, decisions were delayed by over 6 months as against their original completion schedule of 6 months. However owners refuted the above claims by putting the blame on contractor for the nonerection of sheeting/roofing material by contractor and thereby holding the contractors responsible for noncompletion of works inside the building during rainy season. Incidentally, the main contractor had sublet this work of supply and erection of sheeting/roofing to his subcontractor and the sheeting material was not locally available and the same had to be imported from outside the country and there was considerable delay due to this. The owners also blamed contractor for noncompletion of some important structure (overhead water tank) causing delay in commissioning and operation for the said project. On the claim toward the reduction in scope, the owners were of the view that though reduction in the scope was there in some of the items, the contractors have been compensated for the same by additional works.

The contractor however vehemently denied the above allegation and through a letter dated 23rd Dec 1998 stated that even though there was a delay in procuring the imported materials the contractor made an alternative arrangement in place so that the work proceeded unhindered even during the rainy season. The contractor again reiterated that owing to lack of critical inputs in the form of decision, the work was delayed and cited numerous such holds/lack of decision on important issues, last moment changes/ modifications in the drawings/specifications etc.

There was again a round of letters exchanged the following months on the similar issues as pointed above. At last the architect intervened and tried to convene a face- to-face meeting with the contractor and the owner. At one point of time the owners decided to engage the other contractors at site if the existing contractor did not address the owner's issues immediately. Although the joint meetings did help in calming the temper yet the exchange of letters on the same issues continue to surface regularly. Study of all the claims and counter claims of the two parties it appears that the genesis of all the disputes lie in lack of clarity in scope, lack of timely decision, and hold in important decisions and nonavailability of front and inadequate resource mobilization. It appears that the dispute regarding poor quality and bad workmanship can also be attributed to unclear scope and changing requirement. Also due to lack of clarity in scope, the contractor may not have been able to estimate his resource requirement properly and this might have resulted into inadequate resources mobilized at site.

For Project P2 (see Box 4.3), although the commissioning dates were planned, even after initial mobilization of the site by the contractor the work couldn't take off due to the delay in area clearance and the release of drawings. Client and architect couldn't envisage the time taken to get the working drawings approved by the local municipal authorities. The project shows lack of coordination among client, architect, and contractors. There were very few joint meetings and the participants resorted to exchange of letters rather than solving the problems across the table by having face-to-face meetings. Lack of meeting also resulted in widening of conflicts between the project participants. This project also reveals faulty project conceptualization and reluctance in taking decisions by the project participants are sources of project failure by way of time and cost overruns and occurrence of disputes.

Box 4.4 Project P3-Santosh Medical College at Ghaziabad

This project is located in Ghaziabad district of the state of U.P. and consisted of construction of a medical college and hospital building for Maharaja Education Trust. This is a multi-storeyed building. The initial phase of the work was construction of foundation, RCC work of super structure including brickwork. The project was awarded based on negotiation and bids from other parties were not invited. The project duration was 14 months and the contract value was 16 crores.

In this project there was some quality issues raised by the client. The contractor had bought new ply and used for raft and gave good finishing to raft and foundation item. However, in the superstructure, the exposed portion, the contractor used same used ply and the finishes achieved by the old ply shuttering was not acceptable to client. Contractor keeping in mind the economy and the saving was reluctant to procure new ply and continued to use the same old ply. Since the RC work of building was to be left bare without any plaster, and the good finishes through good formwork was known to contractor through drawings and BOQ specifications, the finish provided by contractor by using old ply boards was not acceptable by owner. Also, in this case the owner had engaged a moderate size architectural firm for drawings and design and big contractor for construction. The architect could not keep up the pace of supply of drawings as required by contractor for planning and execution of work. There had been a clear mismatch between the demand of drawings by the contractor and supply of drawings by the architect. Contractor blamed the owner for his incompetency in selecting the architect.

Box 4.5 Project P4-Oriental College at Bhopal

The project was the construction of an engineering college at Bhopal, the capital city of the state of Madhya Pradesh. The construction consisted of multi storeyed buildings and certain utility services for the total cost of Rs. 8 crores. The duration for the project scheduled to be 12 months.

This work was awarded to a big contractor who had already been working in the near vicinity for a mega project of construction of national judicial academy awarded by the Supreme Court of India. Since the contractor had to share all his resources including machinery, tools and plants, and personnel, more attention was always given to the mega project and these engineering college buildings received only the leftovers of the mega project. As a result the quality of the work suffered very badly. Due to engagement of primary

resources in the mega project, this project suffered lack of appropriate resources many times leading to time overrun.
The partial treatment to this project when compared with the mega project was emphasized by both the project manager (of the contractor) and the client to the top management of the contractor firm several times. Yet there was no favorable response forthcoming from the top management. Only after a great loss of time and quality and the notices by the owner, a senior project manager was posted at the site of this project who could get the appropriate resources even from the mega project sites as and when required and saved the image of the company. This is a case of lack of top management support to the site/partial attitude of top management/negligence in carrying out smaller work by big contractors.

Box 4.6 Project P5-Grasim Cement Plant at Chittorgarh
The project consisted of construction of cement plant involving construction of Silos, Preheater building along with other ancillary structures. This cement plant was being constructed at the place called Chittorgarh in the state of Rajasthan. Incidentally the contractor for this cement plant construction had also two more jobs of cement plant construction in the northern region of India in Himachal Pradesh and Madhya Pradesh. Since construction of three cement plants at three locations were going on simultaneously, the crucial resource like slipform materials for the construction of Silos were scarce for three projects.
The project manager of all three sites insisted their top management for procurement and supply of more formwork. Since slipform materials are considered to be special resource and its utility may not be there after completion of these projects, the top management was reluctant to procure additional formwork letting all three project sites suffer in time overrun, quality and finally the cost due to idling of other associated resources.
This is a case of inadequate resources available with the contractor/contractor overloaded with other commitments, as well as the owner's fault of not looking into the concurrent commitments of the contractor while awarding the job.

Box 4.7 Project P6-Construction of Natural Draught Cooling Towers
The project was awarded under competitive bidding to the lowest bidder at an amount of Rs. 17.5 Crores. The contract duration of the project was 24 months. The project consisted of design and construction of two numbers natural draught cooling towers for Rajasthan Atomic Power Project Phase

3&4.The cooling tower was located in a tough terrain and huge amount of blasting in hard rock was required to achieve the founding level of the Raft foundation. While the shell of the natural draught cooling tower was constructed with excellent workmanship with a specialized formwork called Automatic Climbing Formwork used exclusively for Cooling Towers, the other works like Basin works where the water trickles down after cooling was not attended properly. RCC raft was to be laid over the huge amount of rock and sand filling in the basin area. Sand filling in the basin area to reach the required grade before laying the raft was a problem area, as it were to be done in layers and it took enormous amount of time. Contractor, in order to save time did not do a proper filling job which made the clients unhappy and they lodged serious complaint against the project personnel to the top management of contractor. Understanding the seriousness of the problem and realizing the requirement of structural integrity needed for the basin raft, the top management decided to attend to the problem and they sent an experienced project manager to strengthen the project team.
After the induction of an experienced project manager, who could motivate the project staff for good quality work, the quality of work improved and owners got satisfied. This is a case of past experience and motivation skills of PM and the top management support by posting an appropriate PM.

Box 4.8 Project P7: IMCC Project at New Delhi for Delhi Metro Rail Corporation
The scope of this work consisted of construction of a 6.5 km metro twin tunnel system and six underground metro stations for the Delhi Metro Rail Corporation including station air conditioning, tunnel ventilation, station E&M services, lighting, station finishes, and landscaping. The project was bagged by a consortium of DYWIDAG International GmbH (Germany), Larsen & Toubro (India), Samsung Corporation (Korea), IRCON International (India), and Shimizu Corporation (Japan) under the leadership of DYWIDAG International GmbH. This project saw the application of the New Austrian Tunneling method (NATM), earth pressure balance machines (EPBM), and hard rock shield machine. The construction of 300 m long Chawri Bazar Station in the center of Old Delhi for this project is regarded as one of the very challenging works ever done.

Projects P3–P7 (see Box 4.4–4.8) showed lack of quality compliance and there were serious complaints from clients on this aspect. The project managers of these sites were replaced by more experienced and senior managers who could finally finish the work and hand over the projects. This has led to conclude that the causes for the failure of the projects as lack of top management support; incompetent project managers; lack of appropriate resources to carry out the work;

Box 4.9 Project P8-IGI Airport Terminal Building and Extension of Flyover
In order to ease the passenger traffic load, Airport Authority of India decided to construct a new terminal lounge building. They also proposed to widen and extend the existing flyover adjoining the proposed terminal building. Since it was a time bound project, to be completed in 8 months-time, the owner decided to award work on a turnkey basis to a contractor having appropriate expertise. Incidentally the contractor selected for the work was a big construction company and they had opened a new division called Infrastructure projects at their headquarters in the southern part of India. To expand this division and to make a dent in the other parts of the country, the infrastructure division took this project as a challenge, and posted a very experienced project manager at the site. Since, this was a prestigious project and a launching project in the northern part of India, all out support was extended by the top management, besides the work being of turnkey nature, the design team was available in-house and procurement of important machineries, supply of necessary drawings were done ahead of the actual requirement at site. All this finally could save total time of the project and work could be handed over with pride to the client ahead of schedule. Involvement of experienced project manager, top management support, good coordination between project manager's team and head office contributed in the success of the project.

Box 4.10 Project P9- New Nizamuddin Bridge
This bridge is located on NH-24 over river Yamuna in New Delhi. The Ministry of Surface Transport, Government of India awarded this project to Obayashi Corporation. Obayashi awarded part of the job like Construction of Piers on wells, Abutment on either side of the river, 130 nos. 42 m long precast pre-stressed girders, Concrete deck slab, and Asphalt topped carriageway to Indian contractor L&T. The contractual period was from 20th Feb 1996 to 31st Mar 1998, however the contractor M/s L&T completed the job on 12th Feb 1998. The project was hailed a great success both by press and public because of early completion and the quality of workmanship. While this project was able to achieve an extra ordinary pace, a similar adjoining bridge (near ITO) over the same river was running into rough weather every now and then. This bridge (near ITO) was started much earlier than that of New Nizamuddin Bridge and even after a considerable time lapse after contractual period its completion was not in sight. Some of the factors that made New Nizamuddin Bridge a success story are: This being the initial contracting days for the Obayashi Corporation on Indian Soil, they were

eager to leave their footmarks and hence were highly motivated. There was no major change in the drawings issued at the tender stage. They along with the Indian contractor worked out method statement for each and every activity well in advance. Sensing this project as an opportunity the Indian contractor also selected experienced and dedicated staff members including project manager. The staff members with proven track record were brought in from different parts of the country and they gelled together. Most of the staff members toiled very hard and they had to sacrifice their family lives. Round the clock job during most parts of the project were quite normal for this project. While other sites in the vicinity were suffering for want of labor due to outbreak of dengue fever in the capital those days this site was flourishing due to the extra efforts taken by the support staff to prevent the spread of this disease in the labor hutments. Safe work environment was a big draw in pulling the labor force at this site. There was regular coordination meeting between the site staff and regional office staff along with the Obayashi Corporation. Progress of work was regularly monitored and shortfalls if any were promptly addressed to. Sometimes the head office staff also chipped in the coordination meetings to provide additional support. It was with their involvement and support the project saw the employment of 12 cranes, 3 Batching plants and 8 transit mixers during peak days, which was many times higher than a similar project under Indian conditions. Advanced technology called jack down for well sinking also saved time. The Obayashi Corporation also showed their commitment by releasing the payments promptly. The Q/A implementation for the project also ensured improved performance, fewer errors and reworks and hence savings in cost and time.

Box 4.11 Project P10-Hitec City at Hyderabad

This is a prestigious project initiated by the Government of Andhra Pradesh to attract global IT Companies to operate in India. The project was constructed under a joint venture between L&T and Andhra Pradesh Industrial Infrastructure Corporation (APIIC) with 89 and 11 % stake respectively. The project is costing over Rs. 15 billion and is spread over 150 acres. It consisted of a ten storied state-of-the-art building. Since the project was directly monitored by the Chief Minister of Andhra Pradesh, it drew great attention of contractor's top management every time. This led to posting of a very senior and experienced project manager who could draw all types of supports from the headquarters whenever required. The experience of the project manager and the prestige of the project led to good coordination among the project staff in the project site. Coordination between PM's team and head quarter and top management support and motivation to project team also contributed in the success of the project.

Box 4.12 Project P11-Hockey Stadium at Hyderabad

The project was to design and build a twin-court floodlit hockey stadium at Hyderabad and it was to be constructed in 70 days. Hockey field of size 101.4 × 63 m was made up of a sub-base of wet mix macadam, covered by an asphalt layer and topped with synthetic turf. Sophisticated sprinkler system for watering and drainage facility all round the field was also proposed. Each court has an 800 m^2 gallery that can accommodate 500 spectators with many auxiliary units. Apart from architectural civil and structural works the scope included electro-mechanical services and provision of masts for floodlighting. The clients for the project were Sports Authority of Andhra Pradesh, Hyderabad. Since the program for the international sports event was already scheduled and could not be changed, the contractor had mobilized committed team of 3,000 workmen and 50 engineers who could complete the job in time. The success of the project is attributed to commitment of project team and support from top management to achieve the given target.

Box 4.13 Project P12-Construction of Nehru Stadium at Chennai

The Panvel Nadi Viaduct near Ratnagiri in Maharashtra and the Jawaharlal Nehru Stadium at Chennai constructed by ECC has been adjudged the "Most Outstanding Concrete Structures in India for 1994."

Box 4.14 Projects P13 to P16

P13: Construction of Assembly Buildings at Gurgaon for Maruti Udyog Limited

P14: Construction of Moser Baer Factory Building at Greater NOIDA

P15: Construction of Detergent Factory for Proctor & Gamble

P16: Construction of Parliament Library Building for Parliament of India

Three out of these four projects: P13; P14; and P15 were construction of industrial buildings whereas the fourth one: P16 was an institutional building of national importance. The industrial projects were executed for private owners whereas the government of India owned the institutional building.

The investigation of four projects P13–P16 (see Box 4.14), it was observed that these projects met with the desired schedule performance, and commitment shown by the participants involved with these projects was quite commendable.

Interestingly in three out of these four projects the internal assessment committee of contractor awarded the 'team award' to site team recognizing the commitment and team effort shown by every staff member at the project site. The commitment level was so high at these project sites that staff members sometimes even sacrificed their holidays (so scarce in this contracting company) not to mention the long working hours sometimes to the tune of 18–20 h.

Also in all these cases (P13–P16), the owners were very cooperative, and there were no delayed decisions on their part. Owners were very methodical in planning their construction schedule and they were able to anticipate the difficulties lying ahead in their way. The owners helped the contractor out of turn also by releasing advance payments and sometimes modifying the specifications given in tender documents. One of the reasons perhaps for being so cooperative may be that the construction of these buildings was linked to production from these plants. In two of these industrial buildings presence of bonus clause for early completion also the motivating factor that helped in timely completion.

Box 4.15 Project P17-Tarapore Atomic Power Project Unit 3&4

The contract value for the project is over Rs. 1200 Crores. The project consisted of civil, structural, electrical, instrumentation, piping, plumbing and HVAC works for different nuclear structures such as reactor building, turbine building and other ancillary structures. A major portion of the contract is with L&T.

Box 4.16 Project P18—Ramagundam Super Thermal Power

The contractor successfully completed the hydro testing of the 500 MW BHEL Boilers for National Thermal Power Corporation (NTPC) at its Ramagundam Super Thermal Power Project. The contractor completed the hydro testing in a record time of 23 months, which is 5 months and 3 weeks ahead of schedule. Earlier milestone of boiler drum erection was done in 436 days that is 90 days in advance. Some of the remarkable achievements during the course of boiler erection till date are:

Highest tonnage of 395MT erected in any single day

Highest number of 1,006 equivalent joints welded in any single day

Highest tonnage of 1851MT for any single month

Best welder performance in terms of percentage repair of 1.67 in tube welding

The contractor's scope in this project comprises erection, testing and assistance to commissioning of the 500 MW BHEL Boiler. The contractor's site team is well on its way toward ensuring early commissioning of the project.

Project P17 (see Box 4.15), construction of Tarapore Atomic Power Plant 3&4 has been a successful project and the client commended the work by awarding congratulatory letters of appreciation to the contractor's firm for the commitment shown by the project team.

Box 4.17 Project P19-Sri Satya Sai Telgu Ganga Project
The contractor for this project has achieved TEN MILLION SAFE MAN-HOURS. This achievement for the period between July 2002 to June 2003 has been possible because of stringent implementation of safety systems and procedures at site. Involved in this massive project are Sri SatyaSai Central Trust and the Irrigation Department, Govt. of AP. It aims at providing drinking water to Chennai City via the Sai Ganga Canal from the Kandaleru Resorvoir in Andhra Pradesh. The total length of the canal is 48 km from Somasila Reservoir to Kandaleru Reservoir and 152 km from Kandaleru to Tamil Nadu.
The contractor is executing the project with a huge manpower of 4,000 workers and 45 dedicated staff plus a large deployment of equipment. It includes the Slipform Paver, which provides finished profile, friction-free surface and the required compaction.
The major works in contractor's scope comprise:
Improvement work on Kandaleru Reservoir
Improvement to Approach Channel and Sai Ganga Canal
Concrete Lining in Sai Ganga Canal and Somasila-Kandaleru Flood Flow Canal at vulnerable reaches
Construction of Escapes and Regulators
Other miscellaneous Structures and Improvements

4.9 Conclusions from Structured Interviews

As mentioned earlier, a total of 30 structured interviews have been conducted to provide additional insight into the rank order of project success factors, and coordination activities. Explanations from the experts about their rankings offer insights into the rankings obtained through the questionnaire survey. The interviews compliment the questionnaire survey because the interviewees are chosen to represent the experience of the survey population. The interviews started with the researcher briefing about the research project and then finally went on to the subject of the interview. The experts were asked to reveal their preference of the success factors to ensure the overall success of a project. This is done to cross check the rank order of factors affecting success obtained through the mean responses to Q7 of second stage questionnaire as described in Chap. 2. In addition to this experts were also asked to select the three most valuable factors from among the given success factors in order to achieve the schedule, cost, quality, and no-dispute performance criteria.

The rank orders of factors affecting success obtained through the mean values of responses to Q7 of second stage questionnaire and through structured interviews are given in Table 4.9. The format of the Q7 suggests that higher the mean score, higher is its importance while the format of the structured interview suggests that lower the mean score the higher is the importance of the success factor. Spearman's rank correlation techniques was then employed to compare the rank orders for validation of previous results which gave the Spearman's rank correlation coefficient, $\rho = 0.893$ (which is significant at the 0.01 level, 2-tailed) indicating a strong correlation. A strong correlation in the findings of questionnaire survey results and post analysis structured interviews is also observed which also supports the validity of this research finding.

Table 4.9 Comparison of questionnaire survey and structured interviews results

S. No. (1)	Description of success factors (2)	Mean score in questionnaire survey (3)	Rank in questionnaire survey (4)	Mean score in structured interview (5)	Rank in structured interview (6)
1	Project manager's competence	3.74	2	3.20	1
2	Top management support	3.90	1	3.70	2
3	Good coordination between project participants	3.46	4	4.00	3
4	Commitment of all project participants	3.44	5	4.20	4
5	Monitoring and feedback by project participants	3.63	3	5.70	5
6	Availability of trained resources	3.28	8	6.30	6
7	Interaction between project participants-internal	3.44	6	6.40	7
8	Favorable working condition	2.96	11	6.50	8
9	Owners competence	3.30	7	6.50	8
10	Regular budget update	3.14	9	6.70	10
11	Interaction between project participants-external	3.02	10	8.50	11

4.10 Summary and Conclusions

In the second stage questionnaire the respondents were asked to select a project of their own choice which they had executed or with which they were associated. This project was named choice project. Respondents were asked to judge the extent of contribution the success and failure factors in the context of performance of the choice project had on the choice project. Using the performance rating of the choice project as response variable and extent of contribution of various factors as explanatory variables, multinomial logistic regression was applied which led to the following conclusions.

- Extent of contribution of various success or failure factors varies with current level performance ratings of the project. However, six factors (F_3—monitoring and feedback; F_{10}—availability of trained resources; F_{11}—Regular budget update; F_{15}—owner's incompetence; F_{19}—negative attitude of project participants; and F_{20}—faulty project conceptualization) have not been found to cause significant influence on the project outcome.
- None of the factors has been found to have significant influence on all four performance criteria. However, among 11 success factors two factors: F_2—top management support and F_6—owner's competence have been found to be influencing significantly at least three performance criteria. While F_2 contribute in improvement in cost, quality, and no-dispute performance criteria, F_6 contribute in schedule, quality, and no-dispute performances. Similarly, among the failure factors F_{13}—project manager's ignorance and lack of knowledge; and F_{16}—indecisiveness of project participants have been found to be significantly influencing three project performance criteria: schedule, cost, and no-dispute.
- When schedule compliance is the prime objective, seven factors are observed to have significant influence on the schedule outcome. Three factors: F_5—commitment of the project participants; F_6—owner's competence; and F_{12}—conflict among project participants have been found to possess the capability to enhance performance level while the remaining four factors: F_9—coordination among project participants; F_{13}—project manager's ignorance and lack of knowledge; F_{14}—hostile socio economic environment; and F_{16}—indecisiveness of project participants tend to retain the schedule performance at its existing level.
- Factors F_9—coordination among project participants; F_4—favorable working condition; and F_{12}—conflict among project participants are found to be important factors to enhance cost performance of the project. On the other hand, important factors like F_2—top management support; F_5—commitment of project participants; F_{13}—project manager's ignorance; and F_{16}—indecisiveness of project participants tend to keep the cost performance of the project at the same level.
- While no failure factor has emerged out to be significantly affecting the quality performance of project, five success factors have significant influence on the quality performance. The three factors: F_1—project manager's competence; F_2—top management support; and F_8—interaction between project participants-external, contribute significantly in enhancing the project quality performance

from its existing level, the remaining two factors: F_6—owner's competence; and F_7—interaction between project participants-internal, tend to retain the quality performance at the existing level itself. Emergence of project manager's competence and top management support as positive contributor to improve quality reestablishes the findings of quality gurus that management is more responsible to achieve the desired quality in any system.

- A total of six factors have emerged to be significant corresponding to no-dispute criteria. Out of these the factors: F_2—top management support; F_4—favorable working condition; and F_6—owner's competence contribute significantly in avoiding disputes. These factors enhance the probability to avoid dispute only when performance on no-dispute rating is of average nature. None of the success factors considered in the present study seems to have dispute avoiding potential either at low or high dispute ratings.
- Contradictory to the common belief, the factor F_{12}—conflict among project participants is observed to contribute in improvement of the project performance, and this is also in line with the conclusions of some of the past researchers.
- While impact of various factors on project performances has been evaluated in this chapter, methodology to measure and alter the current level of individual factors is yet to be explored. However, the findings given about are expected to give a broad guideline to any professional to select appropriate factor for enhancement or sustenance of the desired level of performance.

The crux of the findings of this chapter has been the emergence of *commitment*, *coordination*, and *competence* as the key factors for achievement of schedule, cost, and quality objectives respectively.

References

Chan, A. P. C., Ho, D. C. K., & Tam, C. M. (2001). Effect of interorganizational teamwork on project outcome. *Journal of Management in Engineering, ASCE, 17*(1), 34–40.

Chan, D. W. M., & Kumaraswamy, M. M. (1997). A comparative study of causes of time overruns in Hong Kong construction projects. *International Journal of Project Management, 15*(1), 55–63.

Cheung, C. C., & Chuah, K. B. (1999). Conflict management styles in Hong Kong industries. *International Journal of Project Management, 17*(6), 393–399.

Cho, C. S., & Gibson, G. E. (2001). Building project scope definition using project definition rating index. *Journal of Architectural Engineering, ASCE, 7*(4), 115–125.

Crowley, L. G., & Karim, A. (1995). Conceptual model of partnering. *Journal of Management in Engineering, ASCE, 11*(5), 33–39.

Deutsch, M. (1994). Constructive conflict resolution: principles, training and research.Conceptual model of partnering. *Journal of Social Issues, 50*(1), 13–32.

Fellows, R., & Liu, A. (1997). *Research methods for construction*. London: Blackwell Science.

Jefferies, M., Gameson, R., & Rowlinson, S. (2002). Critical success factors of the BOOT procurement system: reflections from the Stadium Australia case study. *Engineering Construction and Architectural Management, 9*(4), 352.

Kumaraswamy, M. M. (1998). Consequences of construction conflict: A Hong Kong perspective. *Journal of Management in Engineering, ASCE, 14*(3), 66–74.

Lim, E. H., & Ling, F. Y. Y. (2002). Model for predicting client's contribution to project success. *Engineering Construction and Architectural Management, 9*(5/6), 388–395.

Ling, Y. Y. (2002). Model for predicting performance of architects and engineers. *Journal of Construction Engineering and Mnagement, 128*(5), 446–453.

Songer, A. D., & Molenaar, K. R. (1997). Model for predicting client's contribution to project success project characteristics for successful public-sector design-build. *Journal of Construction Engineering and Management, ASCE, 123*(1), 34–40.

SPSS 9.0.0. SPSS Inc., Headquarters, 233 S. Wacker Drive, 11th floor, Chicago, Illinois, 60606.http://www.spss.com.

Tjosvold, D. (1991). *The conflict positive organization: Stimulate diversity and create unity*. Boston: Addison Wesley.

Whitehead, J.(1998). http://www.personal.ecu.edu/whiteheadj.

Yin, R. K. (1984). *Case study research design and methods*. California: Sage Publications.

Chapter 5
Project Performance Prediction

Abstract A project manager always encounters difficulties in predicting the performance of a construction project. Thus, there is a need to identify the predictor variables used to predict the performance of the construction project. In this chapter the 11 success factors derived earlier have been revisited. Out of these factors, project performance predictors have been identified using artificial neural network. Literature pertaining to performance prediction models has been reviewed and the superiority of ANN in performance prediction is established. Various steps in the ANN applications are clearly explained. The performance prediction models have been derived for all the four project performance criteria: schedule, cost, quality, and no-dispute. The steps to develop a user-interactive model to predict the performance of the construction project based on ANN are also explained. The prediction models may prove to be helpful to the project manager, project team, and top management to predict the performance of the project during its course.

5.1 Introduction

The analysis presented in Chap. 3 produced a number of common factors affecting the schedule, cost, quality, and no-dispute performance criteria, while a few other factors emerged predominantly in only one or more performance criterion. The combination of all the factors across the four performance criteria resulted in 11 success and nine failure factors. The factors are reproduced in Table 5.1.

In Chap. 4, multiple regression analyses were employed and conclusions were drawn by considering the 20 factors as explanatory variables and the effect of these factors on the actual performance as the response variables. The study resulted in factors that needed to be controlled either to enhance or to retain the performance level of the project. Though the critical success and failure factors had been identified, the project managers always encounter difficulties in predicting the

K. N. Jha, *Determinants of Construction Project Success in India*,
Topics in Safety, Risk, Reliability, and Quality 23, DOI: 10.1007/978-94-007-6256-5_5,

Table 5.1 Factors affecting the performance criteria

Factors	Factor description
F_1	Project manager's competence
F_2	Top management support
F_3	Monitoring and feedback by project participants
F_4	Favorable working condition
F_5	Commitment of all project participants
F_6	Owners competence
F_7	Interaction between project participants—internal
F_8	Interaction between project participants—external
F_9	Good coordination between project participants
F_{10}	Availability of trained resources
F_{11}	Regular budget update
F_{12}	Conflict among project participant
F_{13}	PM's ignorance and lack of knowledge
F_{14}	Hostile socio economic environment
F_{15}	Owner's incompetence
F_{16}	Indecisiveness of project participants
F_{17}	Harsh climatic condition at site
F_{18}	Aggressive competition during tendering
F_{19}	Negative attitude of project participants
F_{20}	Faulty project conceptualization

performance of a construction project. Thus, there is a need to identify the predictor variables used to predict the performance of the construction project based on schedule, cost, quality, and dispute and thus this has been the major objective of the present chapter. Another objectives set for the study was to develop a user-interactive model to predict the performance of the construction project, and this also has been discussed in this chapter.

5.2 Performance Prediction Models

There have been several studies to predict the project performance using different project performance criteria. The primary methods of analysis have broadly been the multiple regression analysis or the application of neural network. Koncharandand Sanvido (1998) developed robust models based on multivariate regression analysis to predict unit cost, construction speed, and delivery speed of Design Build (DB) and Design Bid Build (DBB) projects in the US and also identified the specific factors that affect each performance metric. Molenaar and Songer (1998) developed prediction models for public sector projects in the US that relate project characteristics to project success using a multi-attribute regression technique. Performance criteria included budget variance, schedule variance, conformance to expectations, administrative burden, and overall user satisfaction. Since they considered only DB projects, the model had limited applicability.

Chan et al. (2001) constructed models to predict time, cost performance, and overall DB project performance in Hong Kong utilizing data from the 19 projects using the multiple regression technique. Chan et al. (2001) employed factor analysis to reduce 31 attributes into six project success factors: *project team commitment, client's competencies, contractor's competencies, risk and liability assessment, end user's needs,* and *constraints imposed by end users*. These models, however, can only be used to predict the performance of public sector DB projects. Ling et al. (2004) used a multivariate linear regression technique for the development of models to predict construction and delivery speeds.

Chua et al. (1997) constructed a predictive tool to forecast the budget performance of a construction project based on few key management factors. Zin et al. (2006) developed a model to predict the performance of traditional general contract projects based on time, using ANN technique. Ling et al. (2004) also considered ANN technique to construct the models to predict project performance of DB projects based on 11 performance metrics and 65 factors that affect DB project success.

Based on the literature review, a few gaps in knowledge were identified. Most of the studies have dealt with regression models, and they have yielded either low regression coefficients or not provided enough detailed information and thus limiting their usage. The review also shows that the attempts to use linear regression to model the performance of projects with respect to time and cost have not been entirely successful.

Ling et al. (2004) concluded that ANN models have a high degree of predictive ability compared with the linear regression models. Dvir et al. (2006) also concluded that neural networks have better explanatory and prediction powers, which help in deriving relationships among data in a better way compared with the traditional statistical methods. Neural network predictive models have been shown to be capable of forecasting certain project outcomes, without complete information and with a reasonable degree of accuracy (Chua et al. 1997).

It can therefore be concluded that researchers seem to have a general agreement regarding the application of neural networks in developing the prediction models. Even though models based on the studies carried out in developed countries have been developed using ANN; there is a clear need to develop a tool to predict the performance of construction projects in developing countries such as India.

5.3 Research Method

The research method to achieve the objectives of (1) identification of critical factors to predict the schedule performance of construction projects and (2) development of a prediction model for the same has been discussed in this section. Figure 5.1 provides the research method adopted as a schematic diagram and is

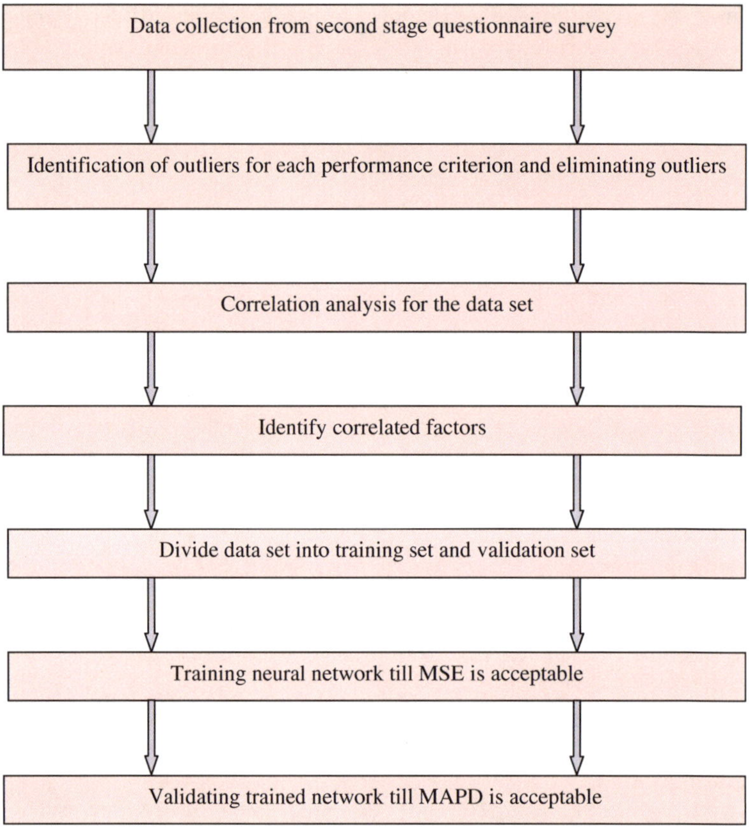

Fig. 5.1 Research method

discussed in the subsequent sections. The process for achieving the objectives stated in this chapter begins with the data preparation and ends with the validation of the prediction model.

5.3.1 Data Preparation

Cheung et al. (2005) emphasized that the development of a model depends on the available data and the procedures of data preparation. The initial step is to list observations with a unique combination of characteristic traits identifiable distinctly from the other observations (outliers). Box plot, a pictorial representation of data distribution, can be used to identify the outliers. Through logical reasoning, the identified outliers can be eliminated from the data set.

5.3.2 Identification of Predictor Variables

The predictor variables that have a high degree of association with the schedule performance can be identified using Spearman's correlation analysis (Ling et al. 2004). Correlation coefficient has the ability to measure the strength of any association between a pair of random variables, and therefore correlation analysis has been chosen. It also measures how a change in one variable affects another variable and vice versa. Here the random variables are treated symmetrically, i.e., the correlation between X_1 and X_2 is the same as the correlation between X_2 and X_1. The correlation is measured on a scale of -1 to $+1$, (-1 represents perfect negative correlation, 0 represents no correlation, and $+1$ represents perfect positive correlation (Soong 2004). A prediction model was developed based on the reduced set of predictor variables.

5.3.3 Prediction Model

In order to validate that the critical factors identified as mentioned in previous section are the key determinants affecting schedule, cost, quality, and dispute performance and also to develop the performance prediction model corresponding to each of the performance criterion, ANN method is employed. ANN modeling was chosen because of its robustness, ability to adapt to unknown data sets, and good learning capability. ANN had been adopted to investigate the factors that affect the different aspects of DB project performance in Singapore construction projects (Ling et al. 2004). A brief introduction to artificial neural network was provided in Chap. 2. In the following sections we provide the details of the proposed neural network model to predict the performance of a construction project.

5.3.4 Network Architecture

Feed-forward neural network architecture was chosen for constructing the ANN model. It consisted of an input layer, an output layer, and hidden layers, if required.

The input layer provides the input data to the network; the size of the input layer, or the number of *neurons (nodes),* indicates the correlated factors for each performance criterion. Gunaydin and Dogan (2004) have emphasized that a numerical scale should be associated with the qualitative data since the network can handle only numeric data. When the hidden layers are used, the number of nodes in the hidden layer may be decided by trial and error. When the input values are received by hidden nodes, they calculate the weighted sum of the inputs and depending upon the transfer function, it gets squashed into a limited range (Edwards 2007). The output layer has one node, which represents the corresponding performance criterion. Table 5.2 gives the description of the sigmoid and linear transfer functions commonly used in MATLAB software.

5.3.5 Training Algorithm

The neural network training minimizes the output error by adjusting network weights and biases. Several learning algorithms have been developed for ANN, but Jain et al. (1996) prefer the back-propagation learning algorithm with feed-forward network architecture for predictions, which is a supervised learning process based on an error correction learning rule. In back-propagation, input and the corresponding output are used to train a network until it can approximate a function and associate input with specific output. In general, properly trained back-propagation networks give reasonable answers when presented with new inputs.

The inputs are sent forward, and then the errors are propagated backwards. The training normally starts with random weights and biases that get adjusted by the algorithm for minimizing errors. The standard back-propagation is a gradient descent algorithm. Other important algorithms are Levenberg–Marquardt (Levenberg 1944 and Marquardt 1963), conjugate gradient algorithm, and resilient back-propagation algorithm. In general, no single algorithm suits all applications; experiments with the algorithms may be done to find the most suitable one for a given application.

5.3.6 Configuration of ANN

After defining the above details, the input–output data are divided into training and validation data sets. Goh (1995) emphasizes the use of about two-thirds of the data for training and the rest for testing and validation. As per the procedure mentioned in Fig. 5.1, the network is trained with a data set that may be stopped once the MSE is acceptable, and for this trained network the validation is carried out. When mean absolute percentage deviation (MAPD) falls within acceptable limits that are application specific, it becomes the validated prediction model.

Table 5.2 Description of transfer functions (Jha and Chockalingam 2011, with permission from Taylor and Francis)

Transfer function	Input range	Output range	Function
Log sigmoid	Plus and minus infinity	0 and 1	$f(x) = \frac{1}{1+e^{-x}}$
Hyperbolic tangent sigmoid	Plus and minus infinity	−1 and +1	$f(x) = \frac{e^x - e^{-x}}{e^x + e^{-x}}$
Linear function	Plus and minus infinity	Plus and minus infinity	$f(x) = x$

5.3.7 *Validation*

The objective of performing validation was to test the capability of the trained network of assessing unknown data sets. Predicted project performance measures derived from the model is compared with actual performance of projects and Table 5.3 gives the performance measures considered to validate the prediction models. When the results of the validation are found acceptable, then the model would be expected to perform well when exposed to new data sets. In Table 5.3, '*n*' stands for the number of predictions.

5.4 Schedule, Cost, Quality, and Dispute Performance Prediction for Construction Projects

The steps mentioned in previous sections are elaborated in the context of present study.

5.4.1 *Data Preparation-Identification of Outliers*

A typical box plot for factor F_1(project manager's competence) for schedule, cost, quality and safety performance is shown in Figs. 5.2, 5.3, 5.4, and 5.5.

In the box plot, the upper and the lower boundaries of the box mark the upper and lower quartiles of the data distribution. The interquartile range (box width) indicates the distance between the 25th quartile and the 75th quartile. The box contains the middle 50 % of the data values. The size of the box indicates the spread of the observation. The whiskers represent the lines drawn from the boundaries of the box to the largest and the smallest values that are not outliers. The 'extreme' outliers represent cases with values of more than 3 box lengths from the boundaries of the box, and the 'mild' outliers represent cases with values between 1.5 and 3 box lengths from the boundaries (Frigge et al. 1989).

Table 5.3 Measures to validate prediction models (Jha and Chockalingam 2011, with permission from Taylor and Francis)

Performance measure	Formula
Percentage deviation (PD)	$PD = \frac{(Actual\,performance - Predicted\,performance)}{(Actual\,performance)} \times 100\,\%$
Mean absolute percentage deviation (MAPD)	$MAPD = \frac{\sum_{i=1}^{n} \lvert PD \rvert}{n}$
Mean square error (MSE)	$MSE = \frac{\sum_{i=1}^{n} (Actual\,performance - Predicted\,performance)^2}{n}$
Root mean square error (RMSE)	$RMSE = \sqrt{MSE}$

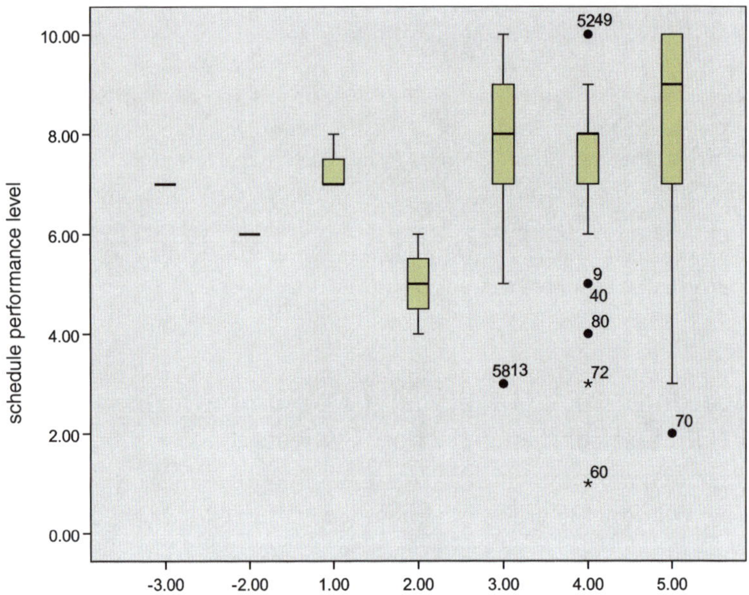

Fig. 5.2 Box plot for the factor 'project manager's competence' for schedule performance (Jha and Chockalingam 2011, with permission from Taylor and Francis)

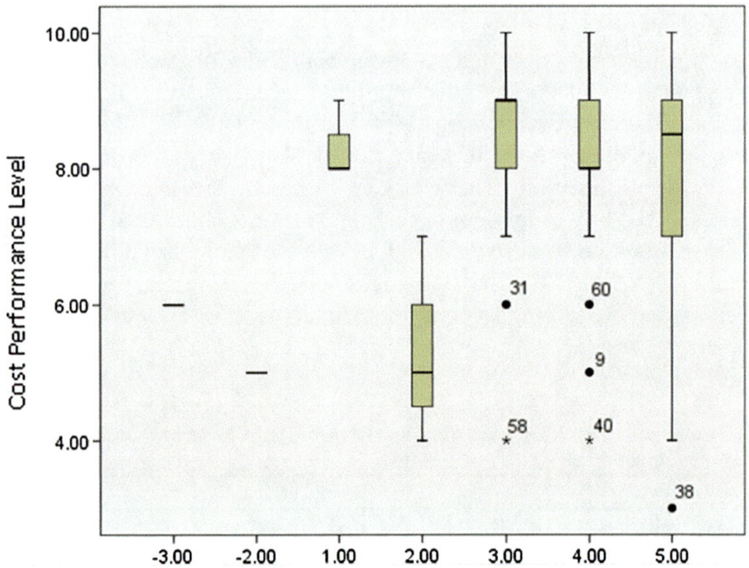

Fig. 5.3 Box plot for the factor 'project manager's competence' for cost performance

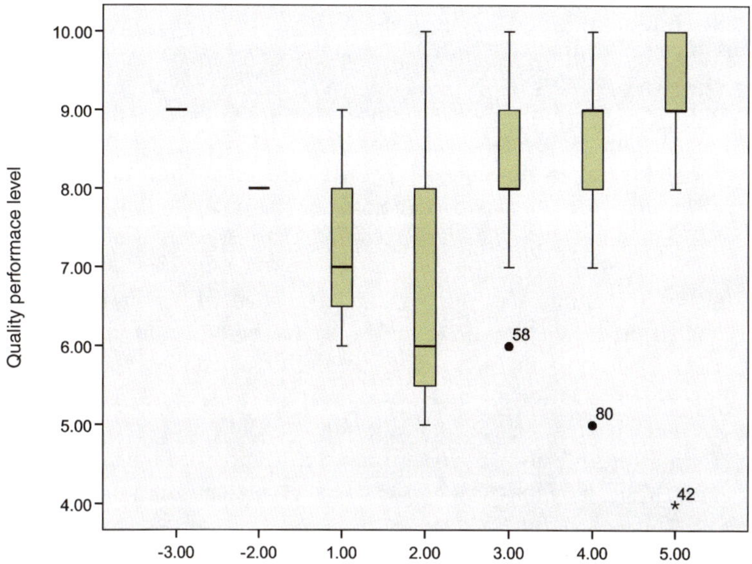

Fig. 5.4 Box plot for the factor 'project manager's competence' for quality performance

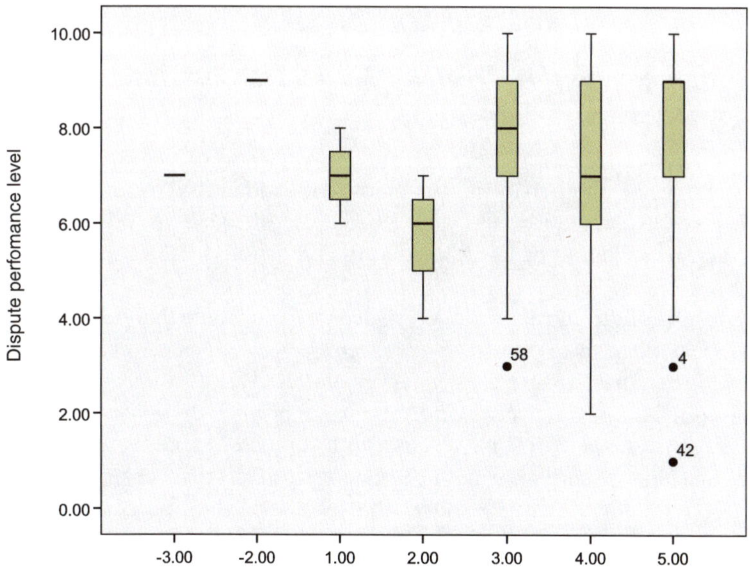

Fig. 5.5 Box plot for factor 'project manager's competence' for dispute performance

In Figs. 5.2, 5.3, 5.4, and 5.5, the cases marked (*) indicate the *extreme* outliers and those marked (o) indicate the 'mild' outliers. In a similar manner, for all the factors such as *top management support, monitoring and feedback by project participants*, etc., the outlier tests were performed and the box plots drawn.

For the schedule performance, the box plots resulted in a common pool of 27 outliers for all 20 factors; however, proper reasoning could only be provided for 15 outliers. Elimination of these 15 outliers from the data set left 76 data sets for further analysis. It may be recalled that a total of 91 responses were obtained in the second stage questionnaire survey.

For the cost performance, the box plots resulted in a common pool of 15 outliers for all the 20 factors, however proper reasoning could be provided for 7 outliers only. Thus, only these 7 outliers were eliminated from the data set which resulted in 84 data sets for the analysis.

For the quality performance, the box plots resulted in a common pool of 14 outliers for all the 20 factors, however proper reasoning could be provided for six outliers only. Thus, only these six outliers were eliminated from the data set which resulted in 85 data sets for the analysis.

For the no-dispute performance, the box plots for all factors resulted in a common pool of 15 outliers, however proper reasoning could be provided for 7 outliers only, hence they are alone eliminated from the data set which resulted in 84 sets for the analysis.

5.4.2 Identification of Predictor Variables-Correlation Analysis

Spearman's correlation analysis was employed to identify the factors that have a high degree of association with the schedule, cost, quality, and dispute performance and which could be considered as the predictor variables for the prediction model.

The results of the correlation analysis led us to conclude that *project manager's competence, monitoring and feedback by project participants, commitment of all project participants, owner's competence, interaction between external project participants,* and *good coordination amongst project participants* significantly affect the schedule performance (factors with sig. 0.000), and the effective control of these factors would prevent the time overrun of the construction projects. Table 5.4 (columns 1 and 2) shows the results of the Spearman's correlation analysis for performance criterion schedule.

For the cost performance criterion, based on the results of the Spearman's correlation analysis, the factors such as *project manager's competence, commitment of all project participants, owner's competence* and *good coordination between project participants* were found to be significantly correlated. These are shown in cols. 3 and 4 of the Table 5.4.

Table 5.4 Spearman correlation results (predictor variables) for schedule, cost, quality, and dispute performance

Schedule performance		Cost performance		Quality performance		Dispute performance	
Factor description	Correlation coefficient	Factor description	Correlation coefficient	Factor description	Correlation coefficient	Factor description	Correlation coefficient
1. F_1-Project manager's competence	0.47	1. F_1-Project manager's competence	0.42	1. F_1-Project manager's competence	0.41	1. F_3- Monitoring and feedback by project participants	0.47
2. F_3-Monitoring and feedback by project participants	0.51	2. F_5-Commitment of all project participants	0.51	2. F_3-Monitoring and feedback by project participants	0.44	2. F_4-Favorable working conditions	0.42
3. F_5-Commitment of all project participants	0.52	3. F_6-Owner's competence	0.45	3. F_5-Commitment of all project participants	0.44	3. F_6-Owner's competence	0.41
4. F_6-Owners competence	0.59	4. F_9-Good coordination between project participants	0.53	4. F_9-Good coordination between project participants	0.50	4. F_9-Good coordination between project participants	0.43
5. F_8-Interaction between project participants—external	0.53			5. F_{10}-Availability of trained resources	0.41	5. F_{11}-Regular budget update	0.36
6. F_9-Good coordination between project participants	0.54						

Spearman's correlation analysis was also used to identify the factors that have a high degree of association with the quality performance. Based on the Spearman's correlation analysis, (see col 5 and 6, Table 5.4) the factors that have a high degree of association with the quality performance were identified. These factors are: *project manager's competence*, *monitoring and feedback by project participants*, *commitment of all project participants*, *good coordination between project participants* and *availability of trained resources* and they were found to be significantly correlated.

The factors that have a high degree of association with the dispute performance and that could be used as predictor variables for the prediction model are shown in the last two columns of Table 5.4. Correlation analysis was preferred as correlation coefficient measures the strength of association between a pair of random variables. The factors: monitoring and feedback by project participants, favorable working conditions, owner's competence, good coordination between project participants and regular budget update are significantly correlated.

5.4.3 Performance Prediction Models

For the performance prediction corresponding to the four performance criterion, ANN models were designed and trained using the MATLAB 7 software. A feed-forward neural network based on back-propagation was applied in the ANN model training, and the sigmoid transfer function was considered for the nodes. Training algorithms and transfer functions were selected based on trial-and-error procedures (Ling and Liu 2004), and experiments were conducted with different performance criteria to obtain the best results.

For the schedule performance, best results were obtained in the fifth trial. In order to accommodate these changes due to number of trials, appropriate programming was done in MATLAB. The Levenberg–Marquardt (Levenberg 1944 and Marquardt 1963) back-propagation algorithm (trainlm in MATLAB) and the transfer function, namely hyperbolic tangent function (tansig in MATLAB) for the neurons in the hidden layers presented quicker convergence and better results during training and validation.

In the Table 5.5, we present a summary of the performance measures such as MAPD, MSE, and RMSE for the models formed by varying the numbers of hidden layers and neurons in each layer corresponding to the four performance criteria namely schedule, cost, quality, and dispute.

For the schedule performance criterion, the 6-3-1 structure gave MAPD 11 % as the lowest considering the single-hidden-layer models. Also the corresponding MSE and RMSE were 0.99 and 1.00, respectively. Among the double-hidden layer models, the 6-1-1-1 structure gave MAPD 11 % as the lowest, and the corresponding MSE and RMSE were 1.01 and 1.00 respectively. These structures are put in a box under schedule column in Table 5.5. The denotation 6-3-1 indicates that the model has 6 nodes in the input layer, 3 nodes in the hidden layer 1 and 1

Table 5.5 Summary of ANN model: network structure and performance measures for schedule, cost, quality, and dispute performances

Schedule performance						Cost performance					
No. of hidden layers	Network structure	MSE (train)	MAPD (%)	MSE (validation)	RMSE (validation)	No. of hidden layers	Network structure	MSE (train)	MAPD (%)	MSE (validation)	RMSE (validation)
1	6-1-1	1.60	13	1.98	1.41	1	4-1-1	0.70	12.378	1.355	1.164
	6-2-1	0.97	12	1.11	1.06		4-2-1	0.72	12.410	1.329	1.153
	6-3-1	0.83	11	0.99	1.00		4-3-1	0.59	12.128	1.494	1.222
	6-4-1	1.38	14	1.90	1.38		4-4-1	0.54	10.379	1.181	1.087
	6-5-1	1.24	14	1.96	1.40						
	6-6-1	1.21	15	2.04	1.43	2	4-1-1-1	1.81	14.131	1.690	1.300
2	6-1-1-1	1.04	11	1.01	1.00		4-2-1-1	1.81	14.053	1.670	1.292
	6-2-1-1	1.31	11	1.09	1.05		4-2-2-1	0.70	12.464	1.367	1.169
	6-2-2-1	1.37	12	1.38	1.18		4-3-1-1	1.56	13.465	1.563	1.250
	6-3-1-1	1.24	13	1.56	1.25		4-3-2-1	0.63	11.900	1.403	1.185
	6-3-2-1	1.35	14	2.16	1.47		4-3-3-1	0.72	12.260	1.413	1.189
	6-3-3-1	1.21	15	1.93	1.39		4-4-1-1	1.80	14.033	1.665	1.290
	6-4-1-1	1.53	13	1.96	1.40		4-4-2-1	0.74	13.663	1.572	1.254
	6-4-2-1	1.21	14	1.87	1.37		4-4-3-1	0.77	14.449	1.723	1.313
	6-4-3-1	1.56	14	2.16	1.47		4-4-4-1	0.73	13.998	1.683	1.297
	6-4-4-1[a]	1.63	15	2.29	1.51						
	6-5-1-1	1.40	14	1.96	1.40						
	6-5-2-1	1.20	13	1.51	1.23						
	6-5-3-1	1.31	16	3.12	1.77						
	6-5-4-1	1.42	14	1.91	1.38						
	6-5-5-1	1.32	14	2.06	1.43						
	6-6-1-1	1.42	13	1.74	1.32						
	6-6-2-1	1.18	16	2.65	1.63						
	6-6-3-1	1.48	17	3.27	1.81						
	6-6-4-1	1.26	14	1.99	1.41						
	6-6-5-1	1.62	16	2.78	1.67						
	6-6-6-1	1.18	14	1.94	1.39						

(continued)

Table 5.5 (continued)

Quality performance						Dispute performance					
No. of hidden layers	Network structure	MSE (train)	MAPD (%)	MSE (validation)	RMSE (validation)	No. of hidden layers	Network structure	MSE (train)	MAPD (%)	MSE (validation)	RMSE (validation)
1	5-1-1	1.14	13.772	2.039	1.428	1	5-1-1	1.76	14.645	2.036	1.427
	5-2-1	0.66	9.321	0.990	0.995		5-2-1	1.25	12.970	1.514	1.230
	5-3-1	0.65	8.681	0.825	0.908		5-3-1	0.92	10.107	0.999	0.999
	5-4-1	0.54	11.540	1.674	1.294		5-4-1	1.36	14.492	1.949	1.396
	5-5-1	0.29	8.044	0.958	0.979		5-5-1	1.94	14.856	2.183	1.478
2	5-1-1-1	0.63	9.347	1.393	1.180	2	5-1-1-1	2.70	15.283	2.162	1.470
	5-2-1-1	1.31	11.788	1.357	1.165		5-2-1-1	1.58	14.450	1.895	1.377
	5-2-2-1	0.57	9.269	0.916	0.957		5-2-2-1	1.85	14.113	1.820	1.349
	5-3-1-1	1.17	11.535	1.323	1.150		5-3-1-1	2.45	14.231	1.919	1.385
	5-3-2-1	1.18	12.268	1.479	1.216		5-3-2-1	2.24	14.223	1.906	1.381
	5-3-3-1	0.67	9.681	1.032	1.016		5-3-3-1	0.98	12.161	1.384	1.177
	5-4-1-1	0.68	11.842	1.933	1.390		5-4-1-1	1.96	13.701	1.690	1.300
	5-4-2-1	1.24	11.705	1.613	1.270		5-4-2-1	1.68	13.087	1.616	1.271
	5-4-3-1	0.82	90439	0.960	0.979		5-4-3-1	2.25	15.031	2.050	1.432
	5-4-4-1	0.56	10.323	1.541	1.241		5-4-4-1	2.15	14.214	1.809	1.345
	5-5-1-1	1.63	12.058	1.411	1.188		5-5-1-1	2.34	13.776	1.915	1.384
	5-5-2-1	1.36	11.755	1.405	1.185		5-5-2-1	0.94	11.570	1.267	1.126
	5-5-3-1	0.62	9.820	1.496	1.223		5-5-3-1	2.06	13.154	1.790	1.338
	5-5-4-1	0.95	11.612	1.279	1.131		5-5-4-1	1.59	13.442	1.622	1.274
	5-5-5-1	1.17	11.089	1.255	1.120		5-5-5-1	1.91	16.425	3.171	1.781

[a] Number of training cycle used in bold faced structures = 1000, in all other structures it was 500

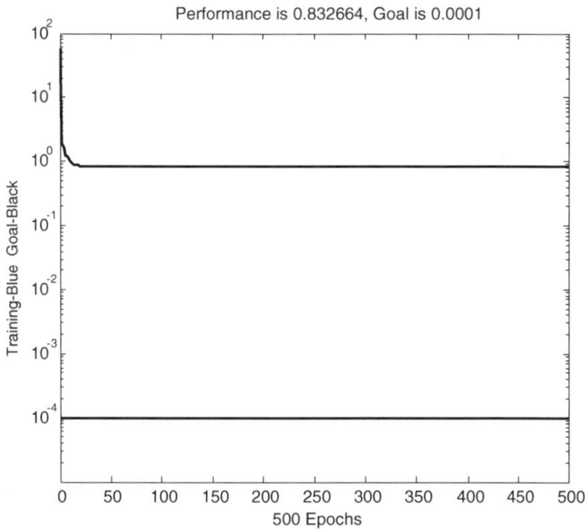

Fig. 5.6 Training curve of ANN 6-3-1 schedule performance model (Jha and Chockalingam 2011, with permission from Taylor and Francis)

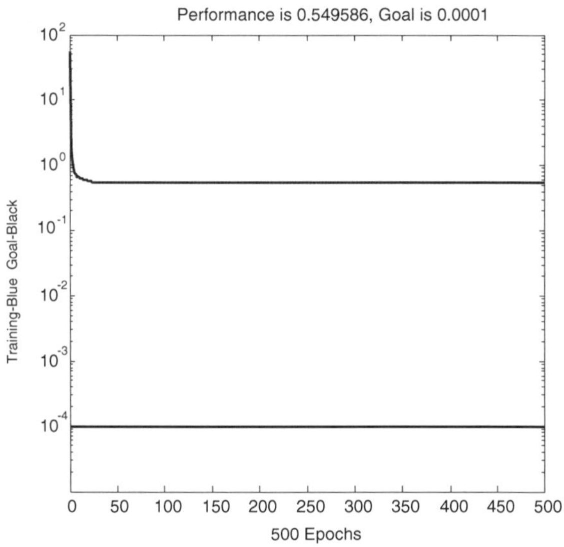

Fig. 5.7 Training curve of ANN 4-4-1 cost performance model

node in the output layer (i.e., schedule performance). In a similar manner, the denotation 6-1-1-1 indicates the model has 6 nodes in the input layer, 1 node in the hidden layer 1, 1 node in the hidden layer 2, and 1 node in the output layer (i.e., schedule performance).

Similar boxes are also shown for other performance criterion. For example in case of cost, quality, and dispute performance criterion, the 4-4-1, 5-5-1, and 5-3-1 structures were found to have the lowest MAPD among all others in single hidden layer network structure respectively. In cases with the two hidden layers, 4-3-2-1, 5-2-2-1, and 5-5-2-1 were found to have the lowest MAPD among all others in

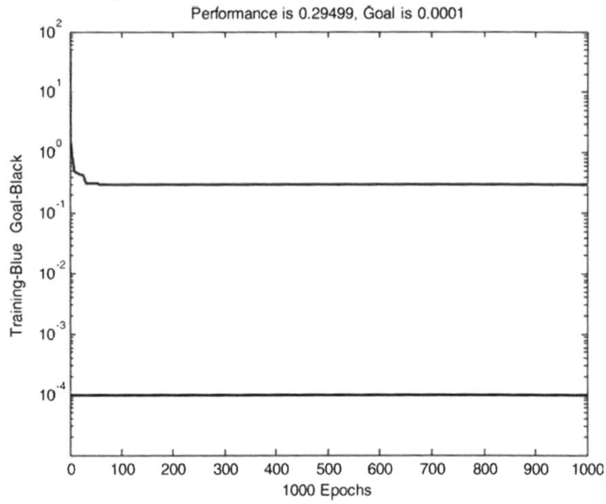

Fig. 5.8 Training curve of ANN 5-5-1 quality performance model

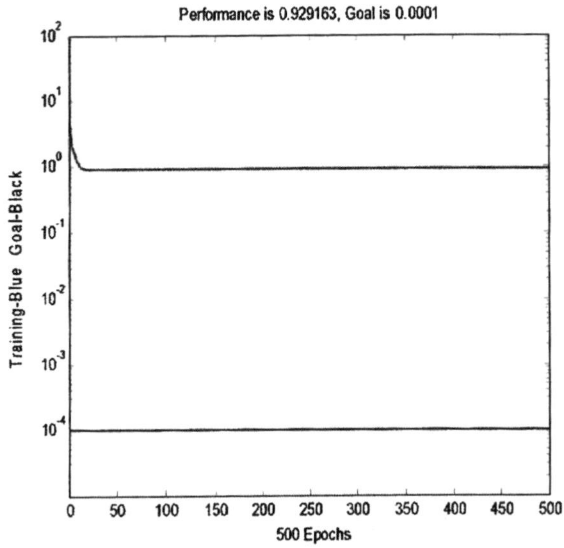

Fig. 5.9 Training curve of ANN 5-3-1 dispute performance model

case of cost, quality, and dispute performances respectively. In all these cases, the RMSE were found to be within reasonable limits and thus acceptable (Ling and Liu 2004). Training curves of the structure mentioned under single hidden layers are shown in Figs. 5.6, 5.7, 5.8, and 5.9 for the four performance criteria.

Subsequently, a comparison of the actual and predicted performances for the two structures with the least MAPD has also been made for the four performance criteria in Table 5.6.

Table 5.6 Comparison of actual and predicted performance for schedule, cost, quality, and dispute

Comparison of 6-3-1 and 6-1-1-1 model						Comparison of 4-4-1 and 4-3-2-1 model					
Schedule performance						Cost performance					
S. No.	Actual performance	ANN 6-3-1 performance	Absolute (PD) (%)	ANN 6-1-1-1 performance	Absolute (PD) (%)	S. No.	Actual performance	ANN 4-4-1 performance	Absolute (PD) (%)	ANN 4-3-2-1 performance	Absolute (PD) (%)
57	10	8.52	15	8.12	18.84	61	9	9.31	3.43	8.71	3.19
58	10	8.66	13	9.10	9.01	62	9	8.83	1.86	8.69	3.48
59	8	7.24	10	7.10	11.23	63	9	8.26	8.23	7.22	19.77
60	5	6.41	28	6.62	32.35	64	5	5.40	8.06	6.80	36.01
61	10	9.40	6	9.37	6.28	65	7	6.81	2.71	7.05	0.67
62	10	9.40	6	9.37	6.28	66	10	8.69	13.12	9.03	9.72
63	9	7.71	14	7.58	15.79	67	10	8.69	13.12	9.03	9.72
64	10	8.41	16	8.93	10.68	68	9	7.87	12.60	7.81	13.25
65	7	8.17	17	8.12	16.06	69	7	8.45	20.64	8.76	25.19
66	9	9.33	4	9.00	0.04	70	9	8.44	6.20	8.44	6.25
67	8	8.73	9	8.62	7.75	71	7	7.37	5.21	7.95	13.52
68	8	7.10	11	6.34	20.71	72	8	8.33	4.13	8.94	11.78
69	8	7.20	10	7.85	1.94	73	7	5.47	21.91	6.88	1.76
70	8	8.05	1	8.23	2.82	74	8	8.71	8.90	8.69	8.60
71	9	8.91	1	8.37	7.06	75	9	7.76	13.73	7.29	19.00
72	8	9.04	13	8.57	7.15	76	9	7.42	17.62	6.94	22.88
73	10	9.32	7	9.23	7.72	77	7	7.14	2.01	7.43	6.16
74	9	8.92	1	8.59	4.53	78	9	8.18	9.10	8.63	4.09
75	9	8.64	4	8.27	8.16	79	8	8.33	4.13	8.94	11.78
76	7	8.79	25	8.37	19.58	80	9	8.69	3.46	9.03	0.31
						81	9	7.98	11.33	8.60	4.41
						82	9	9.31	3.43	8.71	3.19

(continued)

Table 5.6 (continued)

Comparison of 6-3-1 and 6-1-1-1 model						Comparison of 4-4-1 and 4-3-2-1 model					
Schedule performance						Cost performance					
S. No.	Actual performance	ANN 6-3-1 performance	Absolute (PD) (%)	ANN 6-1-1-1 performance	Absolute (PD) (%)	S. No.	Actual performance	ANN 4-4-1 performance	Absolute (PD) (%)	ANN 4-3-2-1 performance	Absolute (PD) (%)
						83	10	7.21	27.89	7.53	24.73
						84	7	8.84	26.32	8.83	26.16

Comparison of 5-5-1 and 5-2-2-1 model							Comparison of 5-3-1 and 5-5-2-1 model					
Quality performance							Dispute performance					
S. No.	Actual performance	ANN 5-5-1 performance	Absolute (PD) (%)	ANN 5-2-2-1 performance	Absolute (PD) (%)	Absolute (PD) (%)	S. No.	Actual performance	ANN 5-3-1 performance	Absolute (PD) (%)	ANN 5-5-2-1 performance	Absolute (PD) (%)
61	9	9.11	1.22	9.24	2.65	19	61	10	9.11	8.89	7.21	27.89
62	8	7.77	2.84	8.91	11.35	9	62	10	9.11	8.89	8.68	13.16
63	10	8.82	11.79	9.13	8.71	11	63	7	7.39	5.63	6.33	9.60
64	10	7.04	29.59	8.00	19.99	32	64	6	7.15	19.09	7.22	20.41
65	8	8.01	0.06	9.13	14.07	6	65	6	3.36	44.03	5.60	6.66
66	6	7.52	25.27	6.40	6.72	6	66	10	9.11	8.89	8.68	13.16
67	6	6.80	13.37	6.84	14.07	16	67	10	9.11	8.89	8.68	13.16
68	10	9.01	9.88	9.25	7.48	11	68	9	7.15	20.61	8.65	3.93
69	10	9.01	9.88	9.25	7.48	16	69	7	7.15	2.08	7.23	3.21
70	9	9.02	0.19	8.29	7.93	0	70	8	9.11	13.89	8.68	8.46
71	7	8.44	20.56	9.07	29.63	8	71	9	9.11	1.23	8.57	4.75
72	10	8.27	17.26	8.98	10.20	21	72	10	9.11	8.89	8.68	13.16
73	8	8.00	0.04	8.59	7.32	2	73	6	7.68	27.99	7.99	33.23
74	9	9.05	0.51	9.17	1.94	3	74	9	9.11	1.23	8.68	3.55

(continued)

Table 5.6 (continued)

Comparison of 5-5-1 and 5-2-2-1 model							Comparison of 5-3-1 and 5-5-2-1 model					
Quality performance							Dispute performance					
S. No.	Actual performance	ANN 5-5-1 performance	Absolute (PD) (%)	ANN 5-2-2-1 performance	Absolute (PD) (%)	Absolute (PD) (%)	S. No.	Actual performance	ANN 5-3-1 performance	Absolute (PD) (%)	ANN 5-5-2-1 performance	Absolute (PD) (%)
75	8	8.98	12.22	9.09	13.67	7	75	8	7.15	10.68	7.11	11.14
76	9	8.48	5.81	7.88	12.46	7	76	7	6.22	11.11	8.67	23.78
77	9	8.26	8.27	7.26	19.29	8	77	8	8.37	4.62	8.88	11.00
78	9	8.27	8.09	8.40	6.71	5	78	9	9.11	1.23	8.69	3.5
79	9	9.53	5.91	8.91	1.01	8	79	9	9.11	1.23	8.68	3.52
80	9	9.63	6.96	9.16	1.78	20	80	8	9.11	13.89	8.69	8.57
81	10	9.11	8.90	9.24	7.62		81	9	9.11	1.23	8.68	3.51
82	9	9.02	0.19	8.93	0.79		82	8	9.11	13.89	7.23	9.69
83	10	9.92	0.88	9.02	9.84		83	9	9.22	2.4	8.58	4.72
84	8	8.00	0.04	8.59	7.32		84	7	7.15	2.08	8.68	23.94
85	9	9.13	1.42	9.15	1.69							

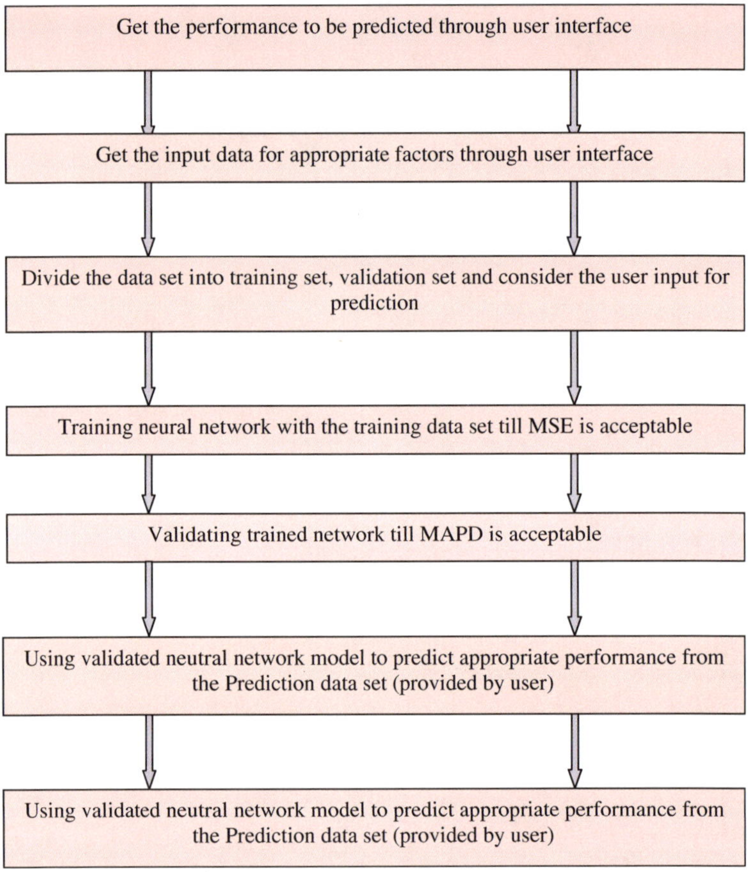

Fig. 5.10 Methodology—user oriented performance prediction model

5.5 User Interface

A user interface has been developed to provide simple access to the developed ANN model and subsequently automate performance prediction. The user interface was developed in GUIDE, the MATLAB graphical user interface development environment that provides a set of tools for creating the graphical user interfaces (GUIs). The GUIDE layout editor was employed to design the buttons, menus etc., and the m-file editor was used to code the callbacks to perform the appropriate functions.

The methodology of the performance prediction model has been given in Fig. 5.10 and the subsequent sections give details of the user interface. With inputs from the user, the project performance is predicted and provided to the user based on the validated prediction models.

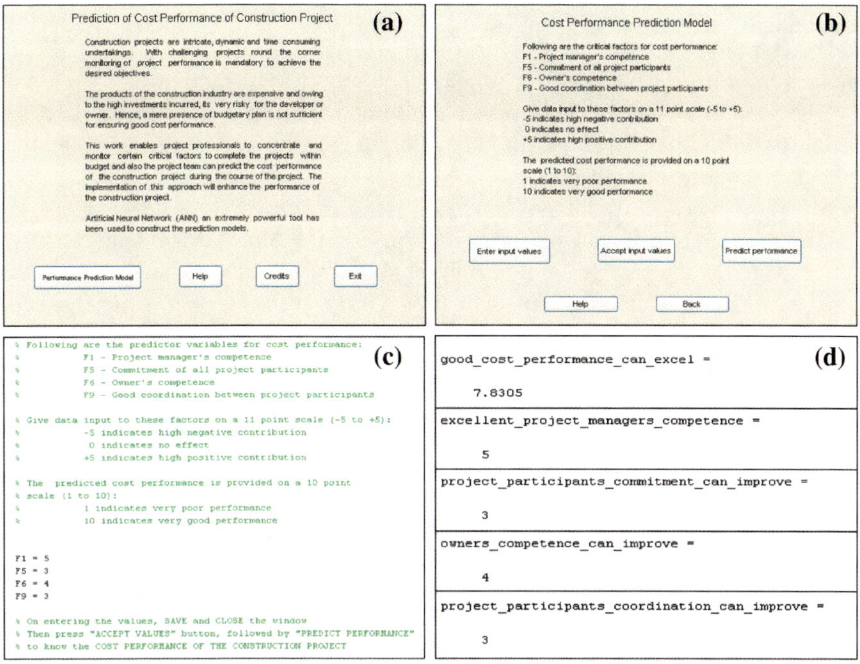

Fig. 5.11 Illustration of user oriented performance prediction model **a** and **b**. User interface screen for cost performance prediction model **c**. Predictor variables input screen **d**. Predicted performance

The main window of the user interface to predict the cost performance is shown in Fig. 5.11a, b. The "enter input values" button in Fig. 5.11b initiates the predictor variables input screen as shown in Fig. 5.11c. The predictor variables F_1, F_5, F_6 and F_9 are accepted as input from the user on a scale of −5 to +5. After data input, the user is to close the input screen and then press "accept input values" button, followed by "predict performance" button in Fig. 5.11b to know the predicted cost performance. The predicted cost performance and the suggestions on the inputs factors appear in the command window as shown in Fig. 5.11d.

The main window of the user interface to predict the project performance is shown in Fig. 5.11a. Figure 5.11b shows the performance prediction model screen. The buttons "schedule", "cost", "quality" and "dispute" initiates the appropriate performance prediction screens.

The cost performance prediction model screen is shown in Fig. 5.11c and the "enter new values" buttons initiates the predictor variables F_1, F_5, F_6, and F_9 (Based on correlation analysis results for cost performance) are accepted as input from the user on a scale of −5 to +5. After data input, the user is to close the input screen and then press "accept input values" button, followed by "predict performance" button in Fig. 5.11c to know the predicted cost performance.

The predicted cost performance and the suggestions on the input factors appear in the command window as shown in Fig. 5.11c in a similar way, the prediction models have been developed for schedule, quality, and dispute also.

The models ensure that the project team and the top management can comfortably predict the project performance during the course of the project and they can also experiment with different scenarios so as to achieve the desired project performance.

The model ensures that the project team and the top management can comfortably predict the cost performance during the course of the project and they can also experiment with different scenarios so as to achieve the desired cost performance.

User-interactive model to predict the cost performance criterion has been developed through correlation studies and with the usage of ANN in MATLAB. The data that has been used for the prediction model development are free from outliers.

5.6 Discussion

Construction projects are dynamic and demanding in nature and so is their management. Atkinson (1999) considered criteria such as organizational benefits, an information system, and benefits for the stakeholder's community in addition to the traditional *iron triangle*. The success of a project may be dependent on a number of factors, but identification of the critical factors helps in the proper allocation of the limited resources of time, labor, and money. Chua et al. (1999) identified budget, schedule, and quality as the critical success factors for different project objectives, using an analytical hierarchy process (AHP). Jha and Iyer (2007) concluded that *commitment, coordination,* and *competence* are the key factors for success of a project, and these need to be managed efficiently to achieve better overall performance.

Dvir et al. (2006) determined critical managerial factors affecting the success of high-tech defense projects by using ANN and regression analysis. Zin et al. (2006) identified 44 factors that affect the traditional general contract success and conducted a questionnaire survey to determine the importance of these factors.

There are not many studies that identify critical factors affecting the schedule, cost, quality, and dispute performance of construction projects in developing countries such as India. Also, there are not many models that are capable of predicting the schedule, cost, quality, and dispute performance of ongoing construction projects. The existing models are either based on the studies carried out in developed countries or restrict themselves to a specific type of construction project.

Development of a prediction model for predicting schedule, cost, quality, and dispute performance of an ongoing construction project using few significant factors, without compromising heavily on the accuracy of the results, has been the major contribution of this study. While some of the significant factors are common for more than one performance criteria, some of the factors are specific only to a single performance criterion. These factors are briefly discussed.

5.6.1 Project Manager's Competence

The project manager's competence has emerged to be a significant factor in the prediction models for schedule, cost, and quality. Zin et al. (2006) also observed that a *project team leader's experience and coordination skills* were significant for better schedule performance in developed countries.

Anderson (1992) classified project manager's managerial attributes as human relationship and leadership skills, technical and administrative experience and also he viewed his position as analogous to that of the CEO if project management is considered as an organizational firm. *Project manager's involvement* and *commitment* plays a major role in project's success and it's affected by the number of projects managed by the project manager at the same time (Chua et al. 1999).

Revelation of project manager's competence as one of the key factors for achieving quality objective also is consistent with the previous studies (Anderson 1992) that project manager's role is instrumental in enhancing the quality performance of the project.

5.6.2 Commitment of Project Participants

This factor has emerged to be significant in schedule and cost prediction. *Commitment,* in general, refers to the willingness of the project participants to work as a team to achieve the same goal. Chan et al. (2001) concluded that a *project team's commitment* is important in achieving better schedule performance. The results of the present study emphasize the *commitment of all project participants.* In other words, the commitments of owners, designers, contractors, subcontractors, consultants, etc., are equally important to achieve better results in India. In fact, in some of the projects considered for validation, commitment level was so high that the project participants worked 18–20 h per day and even sacrificed their holidays at times. In a few cases, contractors were congratulated for their commitment.

In the present scenario, organizations face accelerating growth in areas such as technology, market preference, and pattern of work and legislation in which it can survive only if its employees are effectively trained and fully committed to the growth and development of the company (Tam and Le 2007). Also Tam and Le (2007) emphasized that excellence in quality can be achieved only when the participants are committed.

Chua et al. (1999) observed that project success is highly dependent on *commitment of parties* during the construction phase. Also Kim et al. (2008) have concluded that *commitment of organization* is an important factor related to building projects cost and also it is dependent on the *characteristics of organization* and *participant's competency*.

5.6.3 Monitoring and Feedback by Project Participants

Monitoring and feedback by project participants is important for schedule, quality, and dispute prediction models. In order to ensure schedule, quality, and no-dispute compliance, there should be proper *monitoring and feedback by project participants*, which includes the project team, the project manager, and the owner. The analyses of data sets considered for validation revealed that most projects that achieved better schedule, quality, and dispute performance had an effective monitoring and feedback system.

5.6.4 Owner's Competence

This factor is an important factor in the prediction of schedule, cost, and dispute performances. Most projects that achieved better schedule, cost, and dispute performances had competent owners. *Attitude* and *ability of owner* influences the cost variation of industrial plant projects to a great extent. Also some of the characteristics of a *competent owner* could be his *management ability* and *funding capability*, the *quality of technical specifications and standards* provided and also *possibility of payments delays or repudiation* (Kim et al. 2008). Iyer and Jha (2005) and Jha and Iyer (2006) emphasized that taking *timely decisions, defining scope and nature of work* and *monitoring and feedback* of the progress of the project are also equally important. Chua et al. (1999) have emphasized the importance of a competent owner in the enhancement of project performance. Jin and Ling (2006) observed that partner's incompetence in early stages could impair product quality and trigger litigation among partners. Incompetence could be in the form of inaccurate estimating and inadequate risk provisions, which ends up in strained relationship and poor product quality.

5.6.5 Coordination Between Project Participants

This factor has emerged to be significant across all the performance criteria considered in this study. Construction projects involve multiple participants or stakeholders and to achieve the desired performance and success, proper *coordination* becomes mandatory. Effective *coordination* throughout the project life is facilitated by *interactive processes* that refer to communication, planning, monitoring and control, and project organization (Chua et al. 1999). Also in order to satisfy budget requirements, the *project manager* should *possess good coordinating ability* and *rapport with top management and his project team* (Iyer and Jha 2005).

Chan et al. (2004) also concluded that partnering allows problems to be solved in the shortest time possible and good communication channels i.e., conducting of regular site meetings establishing telephone, fax and e-mail contacts should be adopted to the fullest extent to have performance.

5.6.6 Interaction Between Project Participants

This factor is found to be significant for schedule performance prediction model only. The term 'internal project participants' include the project team, contractors, etc., and the 'external project participants' include the stakeholders like clients, suppliers, etc. In a developing country like India, the results emphasize the importance of *interaction amongst external project participants* in ensuring timely completion of a project. For example, good communication between the project team and the suppliers ensures timely arrival of supplies.

5.6.7 Availability of Trained Resources

This factor has emerged to be significant for quality prediction model only. Employing unskilled labor results in lower output quality that leads to dissatisfaction among owners and most of the low cost labors do not receive proper training (Ling et al. 2007). The present finding that the availability of trained resources is a critical factor to enhance quality performance is also coherent with the results of the above study.

5.6.8 Favorable Working Conditions and Regular Budget Update

These two factors are found to be significant for the dispute prediction model. Favorable working conditions i.e., better climatic conditions avoids disputes among the project participants to a great extent. This is because favorable climatic conditions reduce the chances of facing schedule and cost overrun which in turn reduces the chances of disputes.

The models discussed earlier in the context of developed countries cannot truly replicate the situations prevalent in developing countries, such as India. Factors used in the model may not all be germane, and even their relative importance may vary. Though some of the factors found notable in the study have already been identified as significant in the context of developed countries, certain additional factors also have been identified through the present study. The new model makes use of the factors relevant to developing countries such as India.

The developed models may be helpful to the project manager, project team, and top management to predict the schedule performance of the project during its course. Also, the developed model may help them to select the best strategy among the various strategies available (i.e., each strategy might have different input factors).

5.7 Summary and Conclusions

The significant factors in the schedule performance prediction models are: *project manager's competence, monitoring and feedback by project participants, commitment of all project participants, owner's competence, interaction between external project participants,* and *good coordination between project participants.*

Following are the factors that have been found significant when cost performance is the prime objective: *project manager's competence, commitment of all project participants, owner's competence,* and *good coordination between project participants*. The *commitment* of organization and *attitude & ability of owner* are found to be key factors in past studies to achieve better cost performance (Kim et al. 2008). This is also consistent with findings of Chua et al. (1999) that *project manager's commitment and involvement, owner's and project manager's competency* are significant factors that affect budget performance.

When quality performance is the prime objective, the following are the factors that have been found significant: project manager's competence, monitoring and feedback by project participants, commitment of all project participants, good coordination among project participants, and availability of trained resources. This is consistent with the findings of Ling et al. (2007) and Tam and Le (2007) that adequate training and good coordination are a must for enhanced quality performance. Also the project manager's competence is found to be a key factor in past studies to achieve better quality performance (Anderson 1992).

The factors: monitoring and feedback by project participants, favorable working conditions, owner's competence, good coordination between project participants, and regular budget update are significantly correlated in the case of dispute prediction.

A model has been developed through correlation analysis and the application of ANN to predict the schedule, cost, quality, and dispute performance of a construction project. The ANN models have a feed-forward network based on a back-propagation algorithm, in which the 6-3-1 structure has given the least MAPD of 11 %. The network structure giving the least output in the case of cost, quality, and dispute are 4-4-1, 5-5-1, and 5-3-1 respectively for single hidden layer. In case of two hidden layers, the 6-1-1-1, 4-3-2-1, 5-2-2-1, and 5-5-2-1 structures were found to have the lowest MAPD among all others in case of schedule, cost, quality, and dispute performances respectively. The two structures were also compared for their predictive abilities. The high degree of predictive ability show that the factors identified from correlation analysis are correct and can be used to predict the performance of a construction project in terms of schedule, cost, quality, and dispute.

Development of a performance prediction model for an ongoing construction project using few significant factors, without compromising heavily on the accuracy of the results, has been the major contribution of this study. The developed models may be helpful to the project manager, project team, and top management to predict the schedule performance of the project during its course.

Since the samples drawn belonged mainly to the respondents of large and medium organizations, the results may not be the representative of entire construction industry. Also, inappropriate input data given for the significant factors might give misleading predictions for schedule, cost, quality, and dispute performance; input for significant factors should be given in a responsible manner to get the exact predictions.

References

Anderson, S. D. (1992). Project quality and project managers. *International Journal of Project Management, 10*(3), 138–144.

Atkinson, R. (1999). Project management: Cost, time and quality, two best guesses and a phenomenon, it's time to accept other criteria. *International Journal of Project Management, 17*(6), 337–342.

Chan, A. P. C., Ho, D. C. K., & Tam, C. M. (2001). Design and build project success: multivariate analysis. *Journal of Construction Engineering and Management, ASCE, 127*(2), 93–100.

Chan, A. P. C., Chan, W. M. D., Chiang, Y. H., Tang, B. S., Chan, E. H. W., & Ho, K. S. K. (2004). Exploring critical success factors for partnering in construction projects. *Journal of Construction Engineering and Management*, ASCE, *130*(2), 188–198.

Cheung, S. O., Wong, P. S. P., Fung, A. S. Y., & Coffy, W. V. (2005). Predicting project performance through neural networks. *International Journal of Project Management, 24*(3), 207–215.

Chua, D. K. H., Loh, P. K., Kog, Y. C., & Jaselskis, E. J. (1997). Neural networks for construction project success. *Expert Systems with Applications, 13*(4), 317–328.

Chua, D. K. H., Kog, Y. C., & Loh, P. K. (1999). Critical success factors for different project objectives. *Journal of Construction Engineering and Management*, ASCE, *125*(3), 142–150.

Dvir, D., Ben-David, A., Sadeh, A., & Shenhar, A. J. (2006). Critical managerial factors affecting defense projects success: A comparison between neural network and regression analysis. *Engineering Applications of Artificial Intelligence, 19*(5), 535–543.

Edwards, S. R. (2007). Modelling perceptions of building quality—a neural network approach. *Building and Environment, 42*(7), 2762–2777.

Frigge, M., Hoaglin, D. C., & Iglewicz, B. (1989). Some implementations of the boxplot. *The American Statistician, 43*(1), 50–54.

Goh, A. T. C. (1995). Back propagation neural networks for modeling complex systems. *Artificial Intelligence in Engineering, 9*(3), 143–151.

Gunaydin, H. M., & Dogan, S. Z. (2004). A neural network approach for early cost estimation of structural systems of buildings. *International Journal of Project Management, 22*(7), 595–602.

Iyer, K. C., & Jha, K. N. (2005). Factors affecting cost performance: Evidence from Indian construction projects. *International Journal of Project Management, 23*(4), 283–295.

Jin, X. H., & Ling, F. Y. Y. (2006). Key relationship-based determinants of project performance in China. *Journal of Building Environment, 41*(7), 915–925.

Jain, A. K., Mao, J., & Mohiuddin, K. M. (1996). Artificial neural networks: A tutorial, IEEE (theme feature), 31–44.

Jha, K. N., & Iyer, K. C. (2006). Critical factors affecting quality performance in construction projects. *Total Quality Management, 17*(9), 1155–1170.

Jha, K. N., & Iyer, K. C. (2007). Commitment, coordination, competence and the iron triangle. *International Journal of Project Management, 25*(5), 527–540.

Jha, K. N., & Chockalingam, C. T. (2011). Prediction of schedule performance of Indian construction projects using an artificial neural network, *Construction Management & Economics*, Taylor & Francis, *29*(9),901–911.

Kim, D. Y., Han, S. H., & Kim, H. (2008). Discriminant analysis for predicting ranges of cost variance in international construction projects. *Journal of Construction Engineering and Management, ASCE, 134*(6), 398–410.

Konchar, M., & Sanvido, A. D. (1998). Comparison of US project delivery systems. *Journal of Construction Engineering and Management, ASCE, 124*(6), 435–444.

Levenberg, K. (1944). A method for the solution of certain problems in least squares. *Quarterly of Applied Mathematics, 2*, 164–168.

Ling, F. Y. Y., & Liu, M. (2004). Using neural network to predict performance of design—build projects in Singapore. *Journal of Building and Environment, 39*(10), 1263–1274.

Ling, F. Y. Y., Chan, S. L., Chong, E., & Ee, L. P. (2004). Predicting performance of design—build and design-bid-build projects. *Journal of Construction Engineering and Management*, ASCE, *130*(1), 75–83.

Ling, F. Y. Y., Low, S. P., Wang, S. Q., & Lim, H. H. (2007). Key management practices affecting Singaporean firms' project performance in China. *International Journal of Project Management, 27*(1), 59–71.

Marquardt, D. W. (1963). An algorithm for least-squares estimation of nonlinear parameters. *Journal of the Society of Industrial and Applied Mathematics, 11*(2), 431–441.

Molenaar, K. R., & Songer, A. D. (1998). Model for public sector design build project selection. *Journal of Construction Engineering and Management, ASCE, 124*(6), 467–479.

Soong, T. T. (2004). *Fundamentals of probability and statistics for engineers*. West Sussex: Wiley.

Tam, V. W. Y., & Le, K. N. (2007). Quality improvement in construction industry by using a Vandermonde interpolation technique. *International Journal of Project Management, 25*(8), 815–823.

Zin, M.R., Mansur, S.A., Bakri, A., & Caren, T.C.L. (2006). Predicting the performance of traditional general contract (TGC) projects: a neural network based approach. In: *Proceedings of the 6th Asia-Pacific Structural Engineering and Construction Conference*, (APSEC), 78–86.

Chapter 6
Success Traits for a Construction Project

Abstract From the eleven success factors derived earlier, 'human factors' and 'management actions' have been further analyzed. A hypothesis that 'project success' is influenced by 'success traits' has been formulated. The hypothesized positive inter-relationships between success traits and project success have been tested using the structural equation modeling technique. Various steps taken for achieving the objectives are clearly outlined. The results of the analysis show that hypothesis set for the study holds good. Human factors and management actions play a key role in making the project a success. It can be concluded that trained, committed, competent participant's coordination, with constant monitoring and feedback with regular budget update will influence the successful completion of project.

6.1 Introduction

As mentioned earlier, 20 factors were derived from the preliminary questionnaire survey. Out of these 20 factors, 11 were project success factors.

Success factors pertaining to *human factors* construct and *management actions* construct were considered further for the analysis. As statisticians have suggested that the inclusion of irrelevant variables can result in poor model fit, number of variables should be restricted (Whitehead 1998), so limited attributes from earlier study have been selected and model is analyzed. The CSFs considered under 'human factors' are: project manager's competence, commitment of all project participants, owner's competence, good coordination between project participants, availability of trained resources. The CSFs under 'management action' are monitoring and feedback by project participants and regular budget update.

K. N. Jha, *Determinants of Construction Project Success in India*,
Topics in Safety, Risk, Reliability, and Quality 23, DOI: 10.1007/978-94-007-6256-5_6,

These, 'human factors' and 'management actions', are investigated on schedule, cost, and quality performance, which is considered to be the iron triangle for determining the success.

In this study, it was hypothesized that 'project success' is influenced by 'success traits'. 'Success traits' was defined as a second-ordered construct composed of two latent variables including the human factors and management actions. The chapter utilizes the structural equation modeling technique to test the hypothesized positive inter-relationships between success traits and project success. The hypothesis was tested using structural equation modeling (SEM), a statistical tool.

The main objective of the study is to explore the impact of these success traits on project success.

6.2 Literature Review

The project is at the core of the construction business and project management can be used as a tool to maximize the success of a project (Jaselskis and Ashley 1991).The implementation process is normally uniform across all projects even though each project is unique (PMBOK Guide 2004; Hendrickson and Au 1989). Successful completion of project has many advantages including economic growth and also creates a number of employment opportunities. Additional direct benefits include expanded health facilities, reliable and widespread access to electricity, as well as proper roads, port development, and improved water and sanitation facilities (Orr and Kennedy 2008).

The investigation on construction project success has attracted the interest of many researchers and practitioners (Chua et al. 1999). It is generally accepted that the major goals in a construction project are budget, schedule and quality, although there are other more specific objectives, such as safety consideration, stakeholder's satisfaction and so on depending on the nature of the project, participants and company.

The term critical success factors (CSF) was initially used in the context of information systems and project management by Rockart (1982). Rowlinson (1999) concluded that the CSFs are the fundamental issues that are inherent in the project and they must be maintained in an efficient and effective manner. Chan et al. (2004) developed a conceptual framework on critical success factors (CSFs) and grouped CSFs under five main categories: (1) human-related factors, (2) project-related factors, (3) project procedures, (4) project management actions, and (5) external environment.

The CSFs considered under 'human factors' are: project manager's competence, commitment of all project participants, owner's competence, good coordination between project participants, availability of trained resources. The CSFs under 'management action' are monitoring and feedback by project participants and regular budget update. The project performance is measured based on schedule, cost and quality performance achieved in the project. The project performance on schedule and cost are measured in terms of over/under run as percentage of initial plan, whereas performance on quality is measured based on the compliance/non-compliance of the accepted standards and technical specifications.

Structural equation modeling (SEM) can be thought of as an extension of standardized regression modeling that deals explicitly with poorly measured independent variables. Structural equation models are ideally suited for many of the research issues dealt with in construction engineering and management (Molenaar et al. 2000). SEM is a statistical technique that combines a measurement model (confirmatory factor analysis) and a structural model (regression or path analysis) in a single statistical test (Kline 1998; Mueller 1996; Garver and Mentzer 1999). Theoretically, SEM comprises two types of models: a measurement model and a structural model. The former is concerned with how well the variables measure the latent factors addressing their reliability and validity, and the latter is concerned with modeling the relationships between the latent factors by describing the amount of explained and unexplained variance, which is akin to the system of simultaneous regression models (Wong and Cheung 2005). CFA belongs to the family of SEM techniques as it allows for the assessment of fit between observed data and a priori conceptualized, theoretically grounded model that specifies the hypothesized causal relationships between constructs and their observed indicator variables (Mueller and Hancock 2004).

6.3 Research Method

The research method is schematically presented in Fig. 6.1. The data obtained from second stage questionnaire were analyzed using a software package SPSS 13 and LISERAL 8.8, a structural equation modeling (SEM) tool. The structural equation modeling approach was adopted to understand the causal relations among the various constructs. The first step in SEM is the validation of the measurement model through confirmatory factor analysis (CFA). Figure 6.1 depicts the research method comprising of reliability assessment, exploratory factor analysis, confirmatory factor analysis, and validity tests. These are discussed in the following sections.

6.3.1 Assessment of Reliability

The critical step involved in the development of a measurement scale is the assessment of the reliability of constructs. Different measures serve to analyze the reliability of a construct, Cronbach's alpha coefficient being one of the most often applied. This chapter uses Cronbach's alpha coefficients for assessment of the reliability and validity of the data collection (Nunnally 1978).

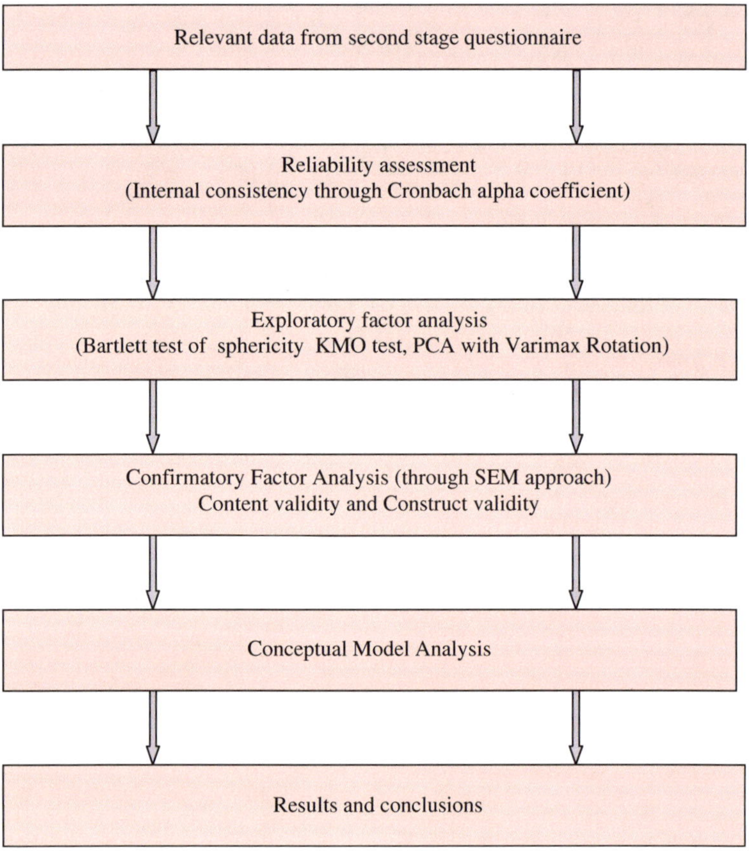

Fig. 6.1 Research method

6.3.2 Exploratory Factor Analysis

The statistical technique of exploratory factor analysis (EFA) was then conducted to check the appropriateness of the proposed grouping of attributes. The appropriateness of data for factor analysis was assessed through the scores for Bartlett test of sphericity and KMO values. The results are compared with standard value and merit of data for acceptability is assessed (Kim and Mueller 1978). Principal component analyses with Varimax rotation were performed to examine the dimensionality of the constructs, and for better interpretability of factor loadings. The communalities and factor loading values are reviewed (Hair et al. 2006; Malhotra 1999).

6.3.3 *Confirmatory Factor Analysis*

Confirmatory factor analysis (CFA) was conducted in order to establish confidence in the measurement model which specifies the posited relations of the observed variables to the underlying constructs. While conducting CFA, construct validity should be satisfied by using content validity and construct validity tests. These are discussed at appropriate places in subsequent sections.

6.3.4 *Conceptual Model and Fit Indices of the Structural Model*

Using the results from preceding section, a conceptual model is hypothesized to test the relationship between the 'success traits' and 'project success'.

6.4 Data Analysis

A hypothesized model covering the 7 attributes for success traits was drawn up to operationally define human and management actions constructs and is shown in Fig. 6.2.

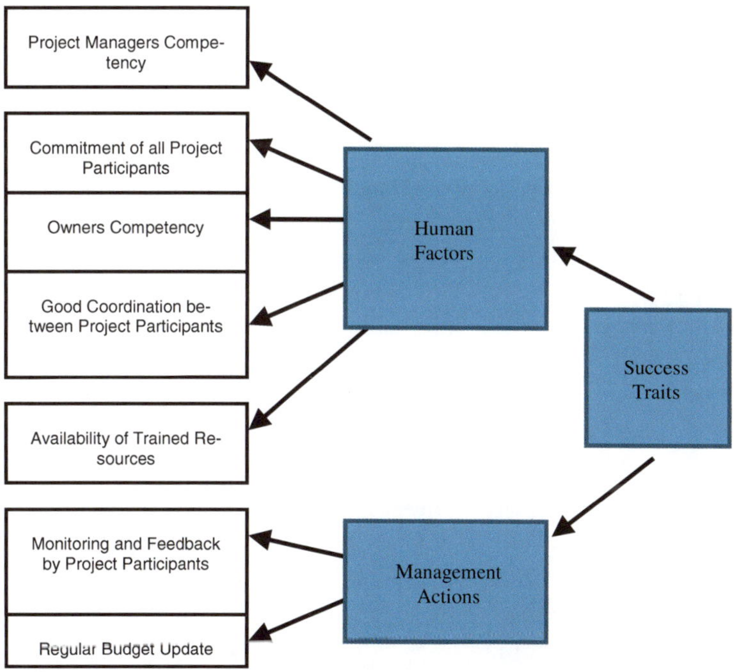

Fig. 6.2 Hypothesized model of success traits

The responses were analyzed to compute their mean scores, standard deviation, skewness and kurtosis, thus ensuring a better understanding of the distribution of each item used in the construct operationalization. The results showed neither outlier nor severely skewed cases, thus increasing the confidence in the contribution of the questionnaire items to the measurement of their respective constructs.

The overall reliability score was found to be 0.795, which is more than 0.7 hence acceptable (Nunnally 1978). The appropriateness of data for exploratory factor analysis was assessed through the KMO value. The KMO value was found to be 0.848 which was greater than 0.6 thus the responses were felt appropriate for factor analysis(Kim and Mueller 1978). Based on Scree plot, the EFA of the 7 attributes extracted two factors accounting for 61.142 % of the total variance. Factor 1 was predominantly accounting for five attributes, initially measuring human related factors and Factor 2 accounting for two attributes, initially measuring management actions.

Further the items in the respective factors were individually subjected to principal component analysis followed by varimax rotation with Kaiser Normalization (for better interpretation). Firstly, the communalities values were reviewed. The communalities indicated the percentage of variance in the variable accounted by extracted factors. The details of the factor loading and communalities are given in Table 6.1. These values were compared with the values recommended by Hair et al. (2006) and Malhotra (1999) found to be appropriate. Suggested minimum acceptable value for reliability is 0.6 (Nunnally 1978; Malhotra 1999). Hence both the dimensions meet the requirement of reliability. This procedure yielded two factors, seven items structure tool for assessing success traits. Accordingly, the hypothesized model of success traits was developed as shown in Fig. 6.2.

The two dimensional structure extracted through exploratory factor analysis approach for the measurement of success traits is validated using confirmatory factor analysis (CFA). Table 6.2 contains the LISREL estimates for the measurement model and the construct correlations. For each success attributes, the t values associated with each of the loadings exceed the critical value 2.576 (at 0.01 significance level). Thus all attributes are significantly related to their specified constructs thereby verifying the posited relationship among indicators and constructs.

6.4.1 Content Validity

While conducting CFA, construct validity should be satisfied by using content validity and empirical validity tests. Once the measurement model is validated, the structural relationships between constructs are estimated (Garver and Mentzer 1999; Anderson and Gerbing 1988). Content validity tests the extent to which a constituent variable belongs to its corresponding construct. Since content validity cannot be tested by using statistical tools, an in-depth literature survey is necessary to keep the researcher's judgment on the right track (Dunn et al. 1994). Accordingly extensive

Table 6.1 Factor structure and loadings of success traits indicators

Success factors	Success traits items	Communalities[a]	Factor loading[b]
Human factors coefficient alpha[c] = 77 % KMO[d] = 0.78	1. Project manager's competence	0.619	0.756
	2. Commitment of all project participants	0.571	0.721
	3. Owners competency	0.449	0.521
	4. Good coordination between project participants	0.751	0.793
	5. Availability of trained resources	0.562	0.623
Management actions coefficient alpha = 60 % KMO = 0.59	1. Monitoring and feedback by project participant	0.619	0.634

[a] Communalities greater than 0.5 is considered appropriate (Hair et al. 2006)
[b] Factor loading greater than 0.5 is acceptable (Hair et al. 2006)
[c] Alpha values of 60 % or higher are considered acceptable (Malhotra 1999)
[d] KMO static value above 0.6 being acceptable (Kim and Muller 1978)

Table 6.2 Construct loadings

Success factors	Success attributes	Factor loading	t-values
Human factors	1. Project manager's competence	0.51	5.11
	2. Commitment of all project participants	0.67	7.14
	3. Owners competency	0.5	4.95
	4. Good coordination between project participants	0.86	9.98
	5. Availability of trained resources	0.68	7.30
Management actions	1. Monitoring and feedback by project participant	0.85	6.73
	2. Regular budget update	0.42	3.96

literature survey was conducted to specify the variables that define constructs. The model was tested in a pilot study administered to industry professionals and academicians and content validity was thus achieved.

6.4.2 Construct Validity

Construct validity comprises of unidimensionality, reliability, convergent, and discriminant validity assessments. Unidimensionality refers to the degree to which constituent variables represent one underlying construct. For testing unidimensionality, a measurement model was specified for each construct and CFA was run for all the constructs. According to Byrne (1998), a comparative fit index (CFI) of 0.90 or above for the model implies that there is a strong evidence of unidimensionality. The CFI value obtained for both the factors in the scale were found to exceed 0.90. This indicated a strong evidence of unidimensionality for the scale.

Reliability assessment was carried out by calculating coefficient alpha separately for each construct. The values obtained were 77 and 60 % respectively for 'human factor' and 'management actions' constructs (see Table 6.1). The results indicated internal consistency among the attributes under each of the constructs.

Convergent validity is the extent to which the construct correlates to corresponding attributes designed to measure the same construct. If the factor loadings are statistically significant, then convergent validity exists. The factor loadings corresponding to the two constructs of the model shown in Table 6.2 indicate that all the factor loadings are significant.

Discriminant validity was assessed through three Chi square comparison models. The comparison of Chi square statistic for Model 0, Model 1, and Model 2 provides support for discriminant validity (Widaman 1985). The Model 0 consists of zero trait and 7 success attributes, Model 1 consists of 1 trait and 7 success attributes, and Model 2 consists of 2 traits and 7 success attributes. The Chi square comparison for Model 0, Model 1, and Model 2 is given in Table 6.3. The Chi square differences are statistically significant, thereby demonstrating discriminant validity.

Another way of assessing construct validity is the goodness-of-fit of the model. A number of fit indices are available, but Marsh et al. (1988) propose that ideal fit indices should have: (1) relative independence of sample size; (2) accuracy and consistency to assess different models; and (3) ease of interpretation aided by a well-defined continuum or pre-set range. Many fit indices do not meet these criteria, because they are adversely affected by sample size (Bentler and Yuan 1999).

In CFA, overall model fit portrays the degree to which the specified attributes represent the hypothesized constructs. For the purpose of confirmatory analysis, we focus only on the measures shown in Table 6.4. The values show that Chi square statistic has a statistical significance and other indices show acceptable fit measures. The Normed fit index (NFI) and adjusted goodness of fit index (AGFI) are 0.97 and 0.95 which are above the recommended level of 0.90, further supporting acceptance of proposed model.

Table 6.3 Chi square comparison

Model No.	Chi square	Degree of freedom	Difference of chi square	Difference of degree of freedom
Model 0 (0 traits and 7 factors)	206.628	21	–	–
Model 1 (1 trait, 7 factors)	14.93	14	Model 0–Model 1 = 206.628–14.93 = 191.698	Model 0–Model 1 = 21−14 = 7
Model 2 (2 traits, 7 factors)	13.39	12	Model 1–Model 2 = 14.93–13.39 = 1.54	Model 1–Model 2 = 14−12 = 2

Table 6.4 Goodness of fit measures for the confirmatory factor analysis

S. No.	Goodness of fit indices	Recommended value	Overall model
1.	χ^2/degree of freedom	<3	8.59/ 13 = 0.66
2.	Goodness of fit index (GFI)	>0.90	0.98
3.	AGFI (Tucker-Lewis index)	>0.90	0.95
4.	NFI	>0.90	0.97
5.	Comparative fit index (CFI)	>0.90	1.0
6.	Root mean-square error of approximation (RMSEA)	<0.1	0.00
7.	Root mean-square residual (RMSR)	<0.05	0.036
8.	Significance level	>0.05	0.80

6.4.3 Conceptual Model and Hypothesis

The results of the preceding sections clearly indicated a two dimensional structure to assess success traits. Using these results a conceptual model is hypothesized to test the relationship between success traits and success of a construction project.

In the proposed model, 'success traits' is considered to be a two-dimensional and second order construct composed of human factors' and 'management actions', and their effect on 'construction project success' is tested. The hypothesized model is shown in Fig. 6.3. The second order approach is recommended by Hair et al. (2006) as it maximizes the interpretability of both the measurement and the structural models. Dark arrow in Fig. 6.3 defines the direction of the influence between two constructs, while light arrows define the dimensions of constructs. Based on the proposed model, the hypothesis that 'success traits' have significant positive influence on construction project success' is tested as below:

Null Hypothesis (H_0): Path coefficient between success traits and construction project success is not significantly different from zero
Alternate Hypothesis (H_a): Success traits have significant positive influence on success
Based on the stated criteria and the suggestions made by various researchers (Garver and Mentzer 1999; Bentlerand Yuan 1999; Jackson 2003) a total of five parameters were selected for validating the hypothesized relationship. These are: (1) the ratio of χ^2 to degree of freedom (2) Goodness of fit (GFI); (3) the non-normed fit index (NNFI); (4) the comparative fit index (CFI); and (5) the root mean squared error of approximation (RMSEA)
The first step involved in validation is to examine the Chi square statistic. The Chi square (χ^2) compares the observed covariance matrix to the one estimated on the assumption that the model being tested is true. But, when the sample size is small, the ratio of χ^2 to degree of freedom (df) is to be examined. Table 6.5 highlights the details of fit measures. The Chi square statistic of the model obtained is 52.74 and the ratio between Chi square and its degree of freedom (χ^2/df) is 1.55 and thus the

model fit is acceptable according to Kline (1998) who specifies a ratio smaller than 3 to be a candidate for an acceptable fit.

The fit of the model was also assessed using goodness-of-fit index (GFI). This is a nonstatistical measure and its value varies from 0 (indicating poor fit) to 1 (indicating perfect fit). In the present case goodness of fit index was found to be 0.91 and thus the model was adequate.

The nonnormed fit index (NNFI) considers a correlation for model complexity Kline (1998). The comparative fit index (CFI) is interpreted in the same way as the NNFI and it represents the relative improvement in fit of the hypothesized model over the null model. In the present case, values obtained for NNFI and CFI exceeds the minimum prescribed value of 0.9. The root mean squared error of approximation (RMSEA) assesses the extent to which the given model approximates the true model. It is an estimate of the discrepancy between the observed and estimated covariance matrices in the population (Hair et al. 2006). In the present case, the value of RMSEA obtained is 0.075 which indicates close fit (Widaman 1985).

The SEM results for the hypothesized model are shown in Fig. 6.3. The relationship between the constructs was hypothesized with a heavy arrow and can be interpreted similar to a regression coefficient that describe the linear relationship between two constructs (Matt and Dean 1993). The coefficients pointing from success traits to the observed variables represent the standardized path coefficients. Larger the coefficient value, more important the variable can be considered as an indicator of success traits. The results of standardized path estimates, t-values and the coefficient of determination (R^2) for the proposed structural model are also shown in Fig. 6.3.

From Fig. 6.3, the relationship between success traits and the two constructs can be expressed as:

$$\text{Success traits} = 0.88\,(\text{management actions}) + 0.98\,(\text{human factors}).$$

The path coefficients in the above equation represent the standardized estimates. The standard error for 'management actions' and 'human factors' are 0.12 and 0.21 respectively and the corresponding t-values associated with the path are 7.47 and 4.57. The *t* test examines the significance of path co-efficient and indicates

Table 6.5 Goodness of fit measures for the structural equation model

S. No.	Goodness of fit indices	Recommended value	Overall model
1.	χ^2/degree of freedom	<3	52.74/ 33 = 1.55
2.	Goodness of fit index (GFI)	>0.90	0.91
3.	Nonnormed fit index (NNFI)	>0.90	0.96
4.	Comparative fit index (CFI)	>0.90	0.97
5.	Root mean-square error of approximation (RMSEA)	<0.1	0.075

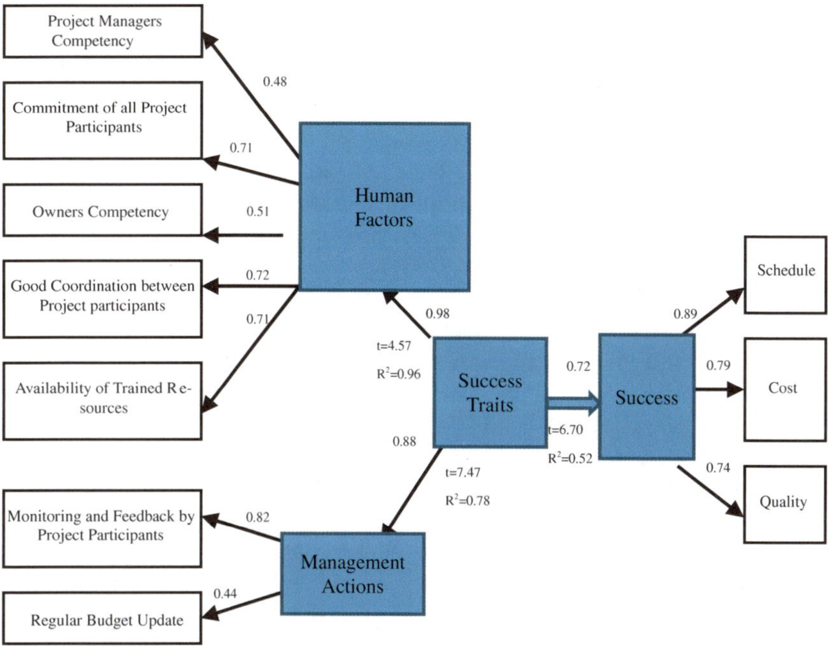

Fig. 6.3 Structural equation modeling results of linkage between success traits and success

whether or not the hypothesized relationship holds. The coefficient of determination R^2 (see Fig. 6.3) values obtained also confirms strong linear relationship among constructs.

All the standardized path coefficients are positive and statistically significant in the desired direction, indicating linkages. The path coefficients being standardized can be taken to be indicative of the relative importance of each construct. The 'human factors' emerges to be the most important success traits. The hypothesis H_1 which assumed that success traits have significant positive impact on success of the projects was found to be supported as the path coefficient (0.72) is significant. The path coefficient marked on this heavy arrow is calculated for a 99 % confidence level and can be interpreted similar to a regression coefficient that describe the linear relationship between two constructs.

6.5 Discussion

The analysis results were found to be acceptable on all the statistical parameters suggested by the statisticians for such study. Thus the hypothesis that 'the success traits defined by 'human factors' and 'management actions' lead to project success' is verified by the analysis results.

The plausibility of project success can be increased if inherent characteristics of the project can be thoroughly understood by trained, well-coordinated, committed and competent management team which also establishes sound monitoring and control system. The outcome of success traits leading to project success identified in this study were found to be consistent with those determined in a previous study using factor analysis with quantitative data and case studies (Jha and Iyer 2007).

The indices suggested by various researchers such as Garverand Mentzer (1999); Bentlerand Yuan (1999); Jackson (2003) and Tabachnick and Fidell (2007) have been used in the study and it is confirmed that there is a considerable influence of 'human factors' and 'management actions' on project success. Human relationship skill, technical knowledge of the subject, belief in team playing spirit and the coordination ability are the key attributes of successful project coordinators (Iyer and Jha 2006). 'Human factors' which is one of the determinants of 'success traits' with a factor loading of 0.98 (Fig. 6.3), depends on the good coordination between project participants, availability of trained resources, competent project manager, and owner and committed participants. Based on their higher factor loadings in Fig. 6.3, it can be stated that 'human factors' are very important for the success of the project. Chan et al. (2001) assert inter organizational teamwork as a major factor in ensuring project success. The reason for the attributes, coordinating ability, commitment and competency of project participants are being given the importance is due to the fact that most of the times their contribution can have far reaching implications on project success. Lim and Ling (2002) emphasize the client's role as an important ingredient in achieving the project success. Taking timely decisions, and regular monitoring and feedback of the progress of the project are some of the characteristics of a competent owner (Iyer and Jha 2006). Munns and Bjeirmi (1996) find that successful project management requires planning with a commitment to complete the project and they further observed that the commitment and support of a parent organization is a vital requirement to project success. Unless the parent organization is willing to commit company resources and provide any necessary administrative support, project management can be very difficult. Commitment basically refers to the willingness of these project participants toward a pooled effort through interaction for making the project a success.

'Management actions' with a factor loading of 0.88 is another major indicator of 'success traits' and in turn impacts project success significantly. 'Management actions' depend on monitoring and feedback and regular budget update. Based on the findings of the study, it can be stated that management in spite of having well-coordinated, committed, competent, and trained participants, should also pay their attention on actions like monitoring and feedback and budget update, to have successful completion of project.

6.6 Summary and Conclusions

The impact of success traits on project success was investigated in this study. SEM was used for the study which apparently has not been used extensively in construction engineering and management research.

According to the model presented in Fig. 6.3, 'success traits' defined by 'human factors' and 'management actions' are the key factors for project success.

A two-step SEM model was set up to measure the four constructs ('human factors', 'management actions', 'success traits', and 'project success') through their constituent variables and to see if the hypothesized relationship holds (Fig. 6.3) good. The analysis was performed as suggested in SEM literature and all the parameters and indices suggested by the statisticians were computed and found to be within acceptable limits. The analysis thus suggested that the hypothesis set for the study holds good with a very strong path coefficient (0.72) as shown in Fig. 6.3.

The construct proposed in our research may help managers to identify the items or dimensions that may lead to successful completion of project. It can be concluded that 'human factors' and 'management actions' play a key role in making the project a success, as seen by the direct link from 'human factors' and 'management actions' related success traits to project success. So, it can be concluded that trained, committed, competent participant's coordination, with constant monitoring and feedback with regular budget update will influence the successful completion of project.

There are certain limitations to this study. The proposed model has been validated by collecting data from large construction senior executives in India only. Also, the self-reported method of data collection has been used thus there may be a possibility of bias playing a role in the final outcome of the study. Nevertheless, this study offers support for the proposed conceptual model and an empirical basis for comparison in future research.

There are several avenues for future work in improving and refining our constructs. First, a larger sample of respondents representing various organizations from different geographical locations can be used to improve the external validity of the proposed construct. Second, it is likely that there may be other dimensions for the individual perceptions of these constructs. These dimensions may be identified and added to the existing construct to improve domain coverage of the constructs.

References

Anderson, J. C., & Gerbing, D. W. (1988). Structural equation modeling in practice: a review and recommended two-step approach. *Psychological Bulletin, 103*(3), 411–423.

Bentler, P. M., & Yuan, K. H. (1999). Structural equation modeling with small samples: test statistics. *Multivariate Behavioral Research, 34*(2), 181–197.

Byrne, B. M. (1998). *Structural equation modelling with LISREL, PRELIS and SIMPLIS: basic concepts, applications and programming*. Mahwah, NJ: Lawrence Erlbaum.

Chan, A. P. C., Ho, D. C. K., & Tam, C. M. (2001). Effect of interorganizational teamwork on project outcome. *Journal of Management and Engineering, 17*(1), 34–40.

Chan, A. P. C., Scott, D., & Chan, P. L. A. (2004). Factor affecting the success of a construction project. *Journal of Construction Engineering and Management, 130*(1), 153–155.

Chua, D. K. H., Kog, Y. C., & Loh, P. K. (1999). Critical success factors for different project objectives. *Journal of Construction Engineering and Management, 125*(3), 142–150.

Dunn, S. C., Seaker, R. F., & Waller, M. A. (1994). Latent variables in business logistics research: scale development and validation. *Journal of Business Logistics, 15*(2), 145–172.

Garver, M. S., & Mentzer, J. T. (1999). Logistics research methods: employing structural equation modeling to test for construct validity. *Journal of Business Logistics, 20*(1), 33–57.

Hair, J. F., Black, W. C., Babin, B. J., Anderson, R. E., & Tatham, R. L. (2006). *Multivariate data analysis* (6th ed.). Upper Saddle River, NJ: Pearson Prentice Hall.

Hendrickson, C., & Au, T. (1989). *Project management for construction: fundamental concepts for owners, engineers, architects and builders*. NY: Prentice Hall.

Iyer, K. C., & Jha, K. N. (2006). Critical factors affecting schedule performance: evidence from Indian construction projects. *Journal of Construction Engineering and Management, 132*(8), 871–881.

Jackson, D. L. (2003). Sample size and number of parameter estimates in maximum likelihood confirmatory factor analysis: a Monte Carlo investigation. *Structural Equation Modelling, 8*(2), 205–223.

Jaselskis, E. J., & Ashley, D. B. (1991). Optimal allocation of project management resources for achieving success. *Journal of Construction Engineering Management, 117*(2), 321–340.

Jha, K. N., & Iyer, K. C. (2007). Commitment, coordination, competency and the iron triangle. *International Journal of Project Management, 25*(5), 527–540.

Kim, J., & Mueller, C. (1978). *Factor analysis*. London: Sage Publications.

Kline, R. B. (1998). *Principles and practice of structural equation modeling*. New York: Guilford Press.

Lim, E. H., & Ling, F. Y. Y. (2002). Model for predicting client's contribution to project success. *Engineering Construction and Architectural Management, 9*(5/6), 388–395.

Malhotra, N. K. (1999). *Marketing research: an applied orientation*. Upper Saddle River: Prentice-Hall.

Marsh, H. W., Balla, J. R., & McDonald, R. P. (1988). Goodness of fit indexes in confirmatory factor analysis: the effect of sample size. *Psychological Bulletin, 103*(3), 391–410.

Matt, G., & Dean, A. (1993). Social support from friends and psychological distress among elderly persons: moderator effects of age. *Journal of Health and Social Behavior, 34*(3), 187–200.

Molenaar, K., Washington, S., & Diekmann, J. (2000). Structural equation model of construction contract dispute potential. *Journal of Construction Engineering and Management, 126*(4), 268–277.

Mueller, R. O. (1996). *Basic principles of structural equation modeling*. New York: Springer.

Mueller, R. O., & Hancock, G. R. (2004). Factor analysis and latent structure, confirmatory. *International Encyclopedia of the Social and Behavioural Sciences*, pp. 5239–5244.

Munns, A. K., & Bjeirmi, B. F. (1996). The role of project management in achieving project success. *International Journal of Project Management, 14*(2), 81–87.

Nunnally, J. (1978). *Psychometric theory* (2nd ed.). New York: McGraw-Hill.

Orr, R., & Kennedy, J. (2008). Highlights of recent trends in global infrastructure: new players and revised game rules. *Transnational Corporations, 17*(1), 95–130.

Rockart, J. F. (1982). The changing role of the information systems executive: a critical success factors perspective. *Sloan Management Review, 23*(3), 3–13.

Rowlinson, S. (1999). Selection criteria. In S. Rowlinson & P. McDermott (Eds.), *Procurement systems: a guide to best practice* (pp. 276–299). London: E and FN Spon.

Tabachnick, B. G., & Fidell, L. S. (2007). *Using multivariate statistics* (5th ed.). Upper Saddle River, NJ: Pearson Education.

Whitehead, J. (1998). Available from: http://www.personal.ecu.edu/whiteheadj.

Widaman, K. F. (1985). Hierarchically nested covariance structure models for multitrait-multimethod data. *Applied Psychological Measurements, 9*(1), 1–26.

Wong, P. S. P., & Cheung, S. O. (2005). Structural equation model of trust and partnering success. *Journal of Management in Engineering, 21*(2), 70–80.

Chapter 7
Project Coordination for Success

Abstract Coordination among project participants is recognized as one of the important success attributes. Coordination activities carried out by a coordinator for achieving day-to-day coordination in a construction project have been identified. Based on the analysis of questionnaire responses the coordination activities have been ranked in descending order of their importance on the four performance criteria: schedule, cost, quality, and no-dispute. The ranking of coordination activities has also been done for the overall success of a construction project and top 20 coordination activities have been identified. The top activities have been evaluated for their criticality in ensuring overall success and the assessment of probabilities of coordination ratings has also been carried out. Further the top coordination activities have also been grouped in fewer factors based on the analysis of second stage questionnaire. The most critical factor to achieve good coordination among project participants has emerged to be 'resource handling', followed by 'planning', 'team-building', and 'contract implementation'.

7.1 Introduction

One of the conclusions in Chap. 4 was that proper coordination among different participants is the most critical factor for any project where cost compliance is of prime importance. Also in the literature, *coordination* has been identified as one of the very important factors by various researchers. There are however, very few systematic studies on coordination in India and abroad. The most recent work on coordination has been by Saram and Ahmed (2001). They have identified a

---

This chapter is based on (a) Critical determinants of project coordination, International Journal of Project Management, 24(4), 2006, with permission from Elsevier, and (b) Ranking and classification of construction coordination activities in Indian projects, Construction Management and Economics, 25(4), 2007, with permission from Taylor and Francis.

K. N. Jha, *Determinants of Construction Project Success in India*,
Topics in Safety, Risk, Reliability, and Quality 23, DOI: 10.1007/978-94-007-6256-5_7,

number of coordination activities performed by the project coordinator in order to achieve day-to-day coordination. Taking lead from this work and the personal interviews with professionals, in the present study a list of 59 coordination activities that are relevant to Indian conditions is prepared and the Q 11 of first stage questionnaire was developed to understand the influence of these coordination activities on various evaluation criteria of the project performance. The broad objectives set for the study are as given below.

- To evaluate the relative importance of the coordination activities in achieving the four-performance criteria: schedule, cost, quality, and no-dispute.
- To study the role of the coordination activities in achieving day-to-day coordination at project sites.

7.2 Coordination: A Key Attribute Recognized

As discussed in the previous section 'coordination among project participants' is identified as one of the important attributes for the success of a project. This has been supported by case studies by a few professionals and researchers, viz., the multi-billion dollar Atlanta Metro Rail Project and World Trade Centre Project, USA etc. (Lammie and Shah 1980; Ruchelman 1980). It is also noticed that lack of coordination among the participants of a project has been responsible for failure of a project such as SCOPE Project, India (India Today 1993) and a host of large building projects in China where cost overrun is recorded to be over 50 % (Wang 2000). Rad (1979) concludes that coordination is the single most important factor in construction of a nuclear power plant when multiple participants are involved. Researchers from other streams viz. Transportation engineering, and water industry have also recognized the importance of coordination among participants (Lam 1991; Grigg 1993). Pinto and Slevin (1988) identify that besides project being technically correct, effective interface of contracting organization with client organization result in success and interfacing effectively with the client organization.

Coordination is also identified as high impact area requiring meticulous and careful handling as far as delays in construction of nuclear power plants is concerned (Rad 1979). Active coordination can minimize, predict and remedy problems caused by design construct lead-time, materials availability, manpower, and equipment availability. It can also minimize problems with design complexity, constructability, productivity and construction methods. Through interviews with experts the researcher proves that coordination is probably the single most important factor in construction of a plant particularly if design and construction contractors and subcontractors are involved.

Through case study of 7 projects, Wang (2000) discusses effects of involvement of foreign designers on coordination issues in Chinese large building projects and on the Chinese construction market and local contractors. While he appreciates the innovations and new concepts in engineering designs brought in by the foreign designers to the local Chinese and breaking the monotony of prevailing Chinese architecture, he is critical about the clear gap between the working patterns or conventions of local Chinese designers and the foreign designers. The Chinese designers emphasize on issue of very detailed drawings and correct estimates to the contractors, whereas not so detailed drawings and correct estimates are provided by foreign designers. This causes problems right from comprehension of drawings to various stages of execution of work including monitoring the physical and financial progress. Also the long distance, ineffective communication technique (lack of e-mail, video conferencing, group work tools etc.) and language problems pose another set of coordination problem, such as inability of designers to visit the sites in regular frequency to give decisions, attend meetings for face to face communication or team building. Hence considerable delay and cost overruns in all projects.

Lammie and Shah (1980) attribute *linking mechanisms* between different levels of project hierarchy for the success of the Atlanta rail transit system project. To manage such a mega project, the entire project is divided into small management groups with specialized task and responsibilities and they are then linked through major integrators viz., owner's general manager, owner's assistant general manager and consultant's project director. The functions of integrators were to ensure that specific decisions, specific actions, and the proper planning perspective were broadcast and coordinated throughout the functional staff.

Lam (1991) discusses success of solving traffic congestion problem in San Diego where over twelve decision making agencies such as local municipal bodies or state agencies called, the land use and transportation agencies, were involved in the process of planning and execution toward solving traffic congestion problem. The objective is achieved successfully primarily through effective coordination among various agencies. He concludes that often lack of coordination among land use and transportation agencies result in severe traffic congestion in upcoming and developing cities. However, if there is a desire on the part of the policymakers to move forward on a project, and willingness on the part of the strongest entities to take the lead in coordinating policies and implementing strategies, such problems can be tackled.

Grigg (1993) observes that coordination in water industry helps in avoiding excessive conflict, wastage of funds and unnecessary damages in the United States. Coordination basically means unifying, harmonizing and integrating different agencies involved in the water industry with multiple objectives. Author cites Sheeran (1976) who considers coordination a principle function of management, ranking along with planning, organizing, directing and controlling. Also coordination has been described as "the all-inclusive management activity" and the process of unifying, harmonizing, and integrating managerial functions, activities, and operations. The author is also of the opinion that coordination happens mostly in the planning process, usually when issues come up in an approval process. In the process, author talks of hydro ecological integration, political integration, geographic integration, functional integration, and disciplinary integration. It is also observed that due to the increase in complexities of a fragmented industry like water industry where a plethora of agencies falling under mainly four heads (water related services providers, regulators, planning-coordinating organizations, and support and assistance organizations) with different objectives are working, there is a greater need of coordination. Author suggests a new paradigm suggesting clear roles for different participants or constituents of the industry. The new paradigm contains (1) recognition of the integrated nature of the water industry; (2) a national water management reporting function; (3) plans and arrangement for the coordination of water management in geographic areas; (4) national water-policy studies; (5) coordination arrangements for water data and research; and (6) a broad array of education and training programs.

Nam and Tatum (1992) find that a high degree of interaction between the design and production functions is closely linked to successful construction innovation. Through a case study of 10 projects (innovative) they conclude that innovative construction projects have certain types of contractual arrangements that encourage the reconciliation of the project participants' conflicting objectives as well as some contractual clauses that foster the exchange of information among the parties involved.

7.3 Coordination Activities in a Construction Project

Saramand Ahmed (2001) point out lack of systematic study and literature in area of construction coordination. They identify 64 construction coordination related activities from texts on duties and responsibilities of various professionals engaged in construction projects. Through questionnaire survey responses these activities are then separately ranked for their order of "importance" and "time consumed" by each of them. Although attempt to estimate the probable time taken for various activities in a quantitative form failed, the response on the qualitative scale prove that the 12 coordination activities ranking as high in "importance" also rank as highly "time consuming". These 12 activities are listed below in the order of importance.

1. Identifying strategic activities and potential delays
2. Maintaining records of all drawings, information, directives, verbal instructions, and documents received from the consultants and the client
3. Maintaining proper relationship with client, consultants, and the contractor
4. Liaison with the client and the consultants
5. Maintaining records of work done outside the contract, variations, day works, and all facts/data necessary to support claims
6. Controlling project finances
7. Establishing and maintaining an effective organizational structure and communication channels

8. Identifying or gathering information on defects, deficiencies, ambiguities and conflicts in drawings and specifications and having them resolved
9. Liaison with specialist consultants, specialist subcontractors, nominated subcontractors etc.
10. Interpreting all contractual commitments and documents
11. Conducting regular meetings and project reviews
12. Analyzing the project performance on time, cost and quality, detecting variances from the schedule/requirements, and dealing with their effects considering time and resource constraints.

7.4 Relative Importance of Project Coordination Activities

Q 11 of first stage questionnaire sought responses on influence of the 59 coordination activities on completion schedule, project cost, project quality, and nonoccurrence of project dispute through a five-point scale. The relative importance or the positive effect of any coordination activity can be gaged by the mean response of the activity. What is important to watch here is that the mean scores of activities vary across the four performance criteria indicating the variation in the degree of positive effect of the coordination activities across the performance criteria. Further, the other interest of the study was to evaluate if the relative importance of various activities in a particular performance criterion remains identical with that of the other criterion. This is done by arranging the activities in a particular rank order and comparing the four rank orders obtained for the four criteria. The scale structure for the question suggests that the lowest mean value of an activity has the largest positive effect on the performance criteria and accordingly can be given the highest rank one and as the mean value increases the rank of the activity in the rank order descends.

Mean values of all the coordination activities and their rank orders under the four performance criteria are summarized criterion-wise in Table 7.1.

The rank orders of the activities in different evaluation criteria suggest that 'regular monitoring of critical path activities for adhering to schedule' is the most important activity when schedule criteria is of prime importance in gauging the project performance, 'monitoring the budget on all activities and taking correcting action' takes supreme importance when cost criterion is considered and similarly 'application of good technical practices' and 'implementation of all contractual commitments' are the most important coordination activities in quality and no-dispute performance criteria respectively. In order to understand better the top five activities of each criterion are summarized in Table 7.2 along with the mean responses.

When schedule achievement is of prime concern, Col 2 of Table 7.2 suggests that monitoring and identification of critical path along with identifying and organizing appropriate resources (rank 5, and 3 respectively) and arranging the

Table 7.1 Rank of coordination activities based on performance criterion

Act No.	Coordination activities	Schedule		Cost		Quality		No-dispute	
		Mean	Rank	Mean	Rank	Mean	Rank	Mean	Rank
1	Implementing all contractual commitments	1.73	20	1.96	12	2.03	16	1.94	1
2	Arranging for timely carrying out of all tests for inspections and approval by the engineer and maintaining records of the same	1.83	33	2.34	39	1.66	3	2.15	3
3	Arranging submission of samples of materials for approval by the engineer	1.99	45	2.49	52	1.71	4	2.41	13
4	Reporting progress reports, resources deployment report etc. as required by the engineer	1.90	40	2.80	58	2.59	50	2.86	36
5	Providing storage space, testing facilities, scaffolding, plant, power, water, illumination, etc. to other agencies as envisaged by the contract	2.07	49	2.57	54	2.63	52	2.78	35
6	Arranging for compliance with site instructions/directives from the engineer and revising programs/ordering material accordingly	1.76	27	2.36	42	2.25	33	2.41	12
7	Applying good technical practices	1.75	23	1.87	7	1.47	1	2.67	28
8	Preparing a project quality plan in line with contract specification	1.90	39	2.24	28	1.49	2	2.35	8
9	Communicating instances of poor quality, dangerous, or adverse incidents/situations to relevant personnel	2.31	53	2.59	55	1.89	8	2.44	15
10	Caring for works of others by making staff and workmen aware of their responsibilities in this regard	2.00	47	2.47	51	2.13	25	2.65	26
11	Coordinating hand over of work areas/service areas (such as plant rooms, service routes, etc.) to other parties	1.88	38	2.45	49	2.72	55	2.40	11
12	Proposing remedial work methods and programs for executing in case of defect or damage	1.85	36	2.15	24	1.96	11	2.44	16
13	Identification of appropriate human resources, materials, and equipments for the project	1.48	5	1.80	5	1.84	5	2.43	14
14	Estimating the optimum resource requirements	1.61	9	1.72	3	2.09	21	2.94	47

(continued)

Table 7.1 (continued)

Act No.	Coordination activities	Schedule		Cost		Quality		No-dispute	
		Mean	Rank	Mean	Rank	Mean	Rank	Mean	Rank
15	Proper assignment of task to the available human resources for the project	1.65	14	1.89	8	1.88	7	2.91	43
16	Organizing resources (manpower, plant, and material) for effective utilization.	1.41	3	1.63	2	2.00	13	2.89	41
17	Arranging technical and behavioral training of human resources.	2.22	51	2.28	31	1.97	12	3.09	55
18	Facilitating payments to own employees and subcontractors	1.75	24	2.49	53	2.21	29	2.63	23
19	Managing the health, safety, and welfare of employees	2.19	50	2.63	57	2.45	45	2.93	46
20	Managing the maintenance and safety of plant and machinery	1.78	29	2.14	23	2.14	26	3.06	51
21	Equipping own men and subcontractors with tools, equipment, and resources	1.82	32	2.28	30	2.06	19	3.20	58
22	Explaining and supporting the work of nominated subcontractors and specialist suppliers	1.91	41	2.41	47	2.12	22	3.12	56
23	Delegation of responsibilities to appropriate project participant	1.75	22	2.33	37	2.06	18	2.86	38
24	Regular follow up of work delegated to project participant	1.59	8	2.38	45	2.02	15	2.77	34
25	Ensuring discipline among all employees	1.73	21	2.25	29	1.91	9	2.69	32
26	Resolving differences/conflicts/ confusion among participants	1.70	19	2.30	34	2.18	27	2.27	6
27	Motivating project participants	1.64	13	2.14	22	2.00	13	2.54	20
28	Developing a team spirit and receiving constructive input from all participants in the project	1.62	10	2.06	16	1.85	6	2.59	21
29	Identifying/gathering information on requirements of all parties and consolidating for use in planning	1.76	44	2.30	33	2.33	39	3.08	53
30	Identification of activities on critical path	1.39	2	1.76	4	2.29	35	3.14	57
31	Communicating project progress, financial and commercial status, plans, schedules, changes, documents, etc., to all relevant participants	1.93	42	2.34	38	2.54	49	3.09	54

(continued)

Table 7.1 (continued)

Act No.	Coordination activities	Schedule		Cost		Quality		No-dispute	
		Mean	Rank	Mean	Rank	Mean	Rank	Mean	Rank
32	Regular monitoring of critical path activities for adhering to schedule	1.30	1	1.90	9	2.36	41	2.98	49
33	Coordinating the purchases, delivery, storage and handling of materials	1.51	6	1.91	10	2.40	42	3.28	59
34	Arranging for kick off meeting and review with all departments asking for date wise schedule for their area of activities.	1.68	17	2.39	46	2.54	48	2.97	48
35	Identifying or gathering information on defects, deficiencies, ambiguities, and conflicts in drawings and specifications and having them resolved	1.84	34	2.13	21	1.93	10	2.68	29
36	Improving/altering/eliminating activities and considering better alternatives that may efficiently meet the project objectives	1.79	30	2.11	17	2.12	23	3.02	50
37	Arranging inputs like drawings, specifications, and technical details on time for execution	1.41	4	1.93	11	2.13	24	2.48	18
38	Providing an organized means for gathering information and compiling records	2.25	52	2.62	56	2.63	53	2.92	45
39	Identifying and gathering information on project work requirements (grouting, openings, making good, etc.) of all relevant parties and coordinating the time and manner of their execution	1.98	26	2.33	36	2.20	28	2.89	40
40	Preparing coordination drawings for freezing sequence of activities and giving a road map of responsibilities to all involved in the project.	1.62	11	2.22	27	2.27	34	2.63	24
41	Agreeing on detailed methods of construction with all the parties involved	1.65	15	2.12	18	2.06	17	2.54	19
42	Coordinating and rescheduling the sequence of onsite work in case of changes in requirement from client side	1.78	28	2.04	14	2.43	43	2.69	31

(continued)

Table 7.1 (continued)

Act No.	Coordination activities	Schedule		Cost		Quality		No-dispute	
		Mean	Rank	Mean	Rank	Mean	Rank	Mean	Rank
43	Interfacing/integrating the work on different subsystems	1.85	35	2.13	20	2.30	36	2.69	32
44	Establishing and maintaining an effective organizational structure and communication channels	1.81	31	2.12	19	2.24	32	2.68	30
45	Conducting regular meetings and project reviews	1.68	18	2.21	25	2.22	31	2.65	25
46	Analyzing the project performances on time, cost and quality, detecting variances from the schedule/requirements and dealing with their effects considering time and resource constraints	1.54	7	1.82	6	2.09	20	2.67	27
47	Monitoring the budget on all activities and taking corrective action	1.86	37	1.51	1	2.49	47	2.90	42
48	Monitoring the overall functioning of each section and department of the project	1.63	12	2.03	13	2.21	30	2.86	37
49	Keeping joint records of all drawings, amendments to contract, directives, correspondences, verbal instructions, and documents received from the project participants (consultants, clients and vendors etc.)	2.06	48	2.45	48	2.43	44	2.45	17
50	Keeping joint records of quantities of work done especially of the work that is to get covered up	2.38	57	2.22	26	2.67	54	2.38	10
51	Keeping joint records of price escalations where the contract has escalation clause.	2.37	56	2.05	15	2.95	58	2.12	2
52	Keeping joint records of owner supplied materials along with their scheduled delivery dates and actual receipt date	2.34	55	2.32	35	2.72	56	2.15	4
53	Keeping joint records of all input cost (viz. labor, material, plant etc.) for nontendered items.	2.61	58	2.29	32	2.80	57	2.18	5
54	Keeping joint records of adverse weather conditions, breakdown time of client supplied equipment etc.	2.32	54	2.46	50	3.09	59	2.28	7

(continued)

Table 7.1 (continued)

Act No.	Coordination activities	Schedule		Cost		Quality		No-dispute	
		Mean	Rank	Mean	Rank	Mean	Rank	Mean	Rank
55	Coordinating with offsite fabricators and their deliveries	1.76	25	2.38	43	2.48	46	2.92	44
56	Acting as liaison with specialist consultants, specialist subcontractors, nominated subcontractors, etc.	1.99	46	2.35	40	2.32	38	2.87	39
57	Maintaining proper relationships with client, consultants and the subcontractor	1.67	16	2.38	44	2.30	37	2.38	9
58	Acting as liaison with the client and the consultants	1.94	43	2.36	41	2.60	51	2.61	22
59	Contacting outside authorities for testing, inspection and approval etc.	2.63	59	2.82	59	2.35	40	3.08	52

required inputs hold the key to successful schedule achievement. For cost as the performance criterion the top ranking activities generally contain resources related activities viz. organizing, estimating, and identification of appropriate resources (ranks 2, 3, and 5). While two of the resource related activities of this criterion are common with that of schedule criterion, the other common activity being 'identification of activities on critical path'. In the quality criterion, activities typically dominate quality compliance activities. It is commonly stated that absence of proper quality plan and systems leads to poor work method causing inconsistent project quality, which sometimes necessitates rework. The lack of quality finally costs by way of extra expenditure on rework and the reputation of the organization. On the 'no-dispute' criterion the top ranked activities suggest adherence to contractual commitment and record keeping are important activities. Record keeping may not improve either schedule or cost, but it does serve as evidence that helps in resolving the dispute faster. Further, it can be observed from Table 7.2 that the mean values of top ranked activities vary from 1.30 to 1.48 in 'schedule' criterion and 1.94 to 2.18 in 'no-dispute' criteria indicating that the level of positive outcome achievable through the coordination activities would be far higher in schedule criterion and quite moderate in no-dispute criterion. From the mean responses of top ranked activities in Table 7.2 it can also be concluded that the maximum achievable positive outcome of coordinating activities varies for the criterion chosen.

In order to understand the contribution of various coordinating activities in a given performance criterion, the activities are classified under three groups based on the mean scores: the first group (with $\mu \leq 1.5$) that shows very large extent of positive effect on the performance criteria; the second group (with $1.5 < \mu \leq 2.5$) which shows large extent of positive effect on the performance criteria; and the

Table 7.2 Summary of top five coordinating activities in different criteria

Rank	Activity description and their mean values for: schedule criterion	Activity description and their mean values for : cost criterion	Activity description and their mean values for: quality criterion	Activity description and their mean values for: no-dispute criterion
1.	Regular monitoring of critical path activities for adhering to schedule (Mean = 1.30)	Monitoring the budget on all activities and taking corrective action (Mean = 1.51)	Applying good technical practices (Mean = 1.47)	Implementing all contractual commitments (Mean = 1.94)
2.	Identification of activities on critical path (Mean = 1.39)	Organizing resources (manpower, plant, and material) for effective utilization) (Mean = 1.63)	Preparing a project quality plan in line with contract specification (Mean = 1.49)	Keeping joint records of price escalations where the contract has escalation clause (Mean = 2.12)
3.	Organizing resources (manpower, plant, and material) for effective utilization) (Mean = 1.41)	Estimate the optimum resource requirements (Mean = 1.72)	Arranging for timely carrying out of all tests for inspections and approval by the engineer and maintaining records of the same (Mean = 1.66)	Arranging for timely carrying out of all tests for inspections and approval by the engineer and maintaining records of the same (Mean = 2.15)
4.	Arranging inputs like drawings, specifications, and technical details on time for execution (Mean = 1.41)	Identification of activities on critical path (Mean = 1.76)	Arranging submission of samples of materials for approval by the engineer (Mean = 1.71)	Keeping joint records of owner supplied materials along with their scheduled delivery dates and actual receipt date (Mean = 2.15)
5.	Identification of appropriate human resources, materials, and equipments for the project (Mean = 1.48)	Identification of appropriate human resources, materials, and equipments for the project (Mean = 1.80)	Identification of appropriate human resources, materials, and equipments for the project (Mean = 1.84)	Keeping joint records of all input cost (viz. labor, material, plant, etc.) for nontendered items. (Mean = 2.18)

third group (with $2.5 < \mu \leq 3.5$) indicating small extent of positive effect on the performance criteria. The number of activities falling in different groups is presented in Table 7.3. Going through this table it is observed that while 18 coordination activities have very large effect in influencing the schedule performance, only the top two and three activities have very large effect in influencing the cost

Table 7.3 Number of activities in different groups

Range of mean values	Number of activities emerged when Schedule is the criterion	Number of activities emerged when Cost is the criterion	Number of activities emerged when Quality is the criterion	Number of activities emerged when No-dispute the criterion
$\mu \leq 1.5$ (very large effect)	Top 18 activities	Top two activities	Top three activities	None
$1.5 < \mu \leq 2.5$ (large effect)	Remaining 41 activities	Activities from Rank 3 to Rank 57	Activities from Rank 4 to Rank 57	Top 35 activities
$2.5 < \mu \leq 3.5$ (small effect)	None	Remaining two activities	Remaining two activities	The remaining 24 activities

and quality performances respectively. It can also be observed from Table 7.3 that none of the 59 coordination activities has very large effect on dispute prevention.

7.5 Differences in Perception on Various Coordinating Activities

As discussed earlier the respondents are placed in two broad groups: owner and contractor. In order to understand if there exists any difference in perception between these two groups of respondents on the contribution of the coordination activities on achieving the project performance objectives or their relative positioning (ranks in the rank order), Analysis of variance (ANOVA) is carried out between the means of responses of owners and contractors. The difference in opinion for the different activities under the four-performance criteria has been discussed in the following paragraphs.

7.5.1 Schedule Criterion

ANOVA results point out differences in perception between owner and contractor in only eight (8) activities out of 59 activities. These eight activities along with their mean values under two groups, their original ranks, F-statistics and the significance level (α) at which hypothesis of equality of mean values across different groups could be rejected are summarized in Table 7.4. A lower value of significance level indicates that null hypothesis of equality of mean can be rejected, i.e., mean values between the two groups vary significantly. For example, the significance level values less than 0.05 would indicate that there is a difference of opinion among different groups on the parameter in question at 95 % confidence level. Subsequent to these analyses, the results were discussed with a few professionals for getting greater insight of the results.

From activity serial number 1–4 in Table 7.4, it can be seen that their mean values are less than 1.5 in owner response and they have got very high ranks, while in contractor response the mean values are greater than 1.5 and ranks are also lower. The description of the first activity, 'estimating the optimum resource requirements', if observed carefully, points out the owner's concern for the estimate of optimum resource that should always be there to feed for the work as and when required. The second activity can be said to indicate that owner feels that if available human resources are utilized appropriately more productivity can be achieved that would earn time for other works and the third activity indicates his concern to sort out all project related bottlenecks in meetings so that everybody is aware of his/her responsibility and work goes smoother and faster. Similarly, the fourth activity with rank 9 in owner ranking indicates the preparedness of owner or his engineer with the drawing details to facilitate plan and complete the work by

Table 7.4 Summary of ANOVA results of important activities between contractor and owner responses on Schedule Criterion (Data arranged in the increasing mean value in owner response)

S. No.	Coordination Activities	Mean of owner response	Rank of owner response	Mean of contractor response	Rank of contractor response	F-value	α
1.	Estimating the optimum resource requirements	1.45	6	1.77	17	2.85	0.10
2.	Proper assignment of task to the available human resources for the project	1.47	7	1.81	22	3.66	0.06
3.	Conducting regular meetings and project reviews	1.47	8	1.92	37	6.51	0.01
4.	Preparing coordination drawings for freezing sequence of activities and giving a road map of responsibilities to all involved in the project	1.49	9	1.84	24	3.31	0.07
5.	Identification of activities on critical path	1.50	10	1.20	1	3.02	0.09
6.	Preparing a project quality plan in line with contract specification	1.69	23	2.12	47	2.84	0.10
7.	Arranging technical and behavioral training of human resources	1.85	37	2.54	56	10.37	0.00
8.	Providing storage space, testing facilities, scaffolding, plant, power, water, illumination, etc. to other agencies as envisaged by the contract	1.86	38	2.40	53	3.88	0.05

contractor on time. High ranking of these activities by contractors may be said to show over confidence in handling these activities or some other activities are considered to be more important.

For example, the fifth activity, 'identification of activities on critical path' is ranked 1st by contractors while it is ranked 10th by owners. Here contractor may probably be of the view that once activities of critical path are identified and taken care of project would go smoothly, while owner's concern is that if contractor cares for critical activities of the original schedule network, and does not care much for the noncritical activities, chances of slippage may occur and these noncritical activities may enter into the critical path and hamper the schedule. Owners, through their responses, may like to caution that contractors should be equally careful in handling noncritical activities as that of critical activities.

Regarding the sixth activity, 'Preparing a project quality plan in line with contract specification' with the rank 23 in owner response and rank 47 in contractor response may be said to convey owners' advice that 'prevention is better than cure'. Quality plan should not be considered as impediment in progress of work. Proper quality plan and adherence to it will prevent loss of time for redoing or correcting the defective work. Low rank in contractor response may either indicate that they are too confident about their quality and do not attribute importance to this, or they think that they can get away even with defective work whenever it arises. In the remaining two activities, ranks are relatively low in both responses, so it is difficult to offer reasons for differences in perception.

7.5.2 Cost Criterion

ANOVA results pointed out differences in perception between owner and contractor in only two (2) activities out of 59 activities. These two activities along with their mean values under two groups, their original ranks, F-statistics and the significance level (α) at which hypothesis of equality of mean values across different groups could be rejected are summarized in Table 7.5.

From Table 7.5 it can be seen that activity, 'Conducting regular meetings and project reviews' has received a low rank in the contractor response (rank 47) while

Table 7.5 Summary of ANOVA results of important activities between contractor and owner responses on Cost criterion (Data arranged in the increasing mean value in owner response)

S. No.	Coordination activities	Mean of owner response	Rank of owner response	Mean of contractor response	Rank of contractor response	F-value	α
1.	Conducting regular meetings and project reviews	2.06	14	2.50	47	2.85	0.10
2.	Arranging inputs like drawings, specifications, and technical details on time for execution	2.11	25	1.58	4	4.71	0.03

it has secured a very high rank (rank 14) in owner responses. Generally in the review meetings owners' overall project objectives are emphasized heavily and the contractors are normally reminded of their shortfalls in different areas making it an unpleasant encounter for the contractor. Contractors feel this is more of an obligation and consider it to be waste of time and probably in the response too they project this activity to be one of the least important one. On the contrary, owners feel that most of the bottleneck of the project can be sorted out and any pending decision can be quickly sought in such meetings. Hence they give high importance to such meetings. So the differences observed in the responses are valid.

Similarly for a contractor 'getting the required inputs like drawings/specifications etc.' (rank 4) are very important for they have to plan the activities and resources in an optimal way to minimize cost of the project, whereas for an owner this may not be of that priority (rank 25).

Though the differences between contractor and owner responses may be observed only in two activities out of 59 activities it cannot be ignored as they actually represent the expected output from other side. As it is generally said, in order to have good coordination between team participants, one must respect the other's expectations and act accordingly.

7.5.3 Quality Criterion

ANOVA results point out differences in perception between owner and contractor in seven (7) activities out of 59 activities. These seven activities along with their mean values under two groups, their original ranks, F-statistics and the significance level (α) at which hypothesis of equality of mean values across different groups could be rejected are summarized in Table 7.6.

First activity in Table 7.6 suggests owner's concern for identifying problem areas in the drawings and specification. Owners view this as their prime responsibility since clearing this at the very first instance paves the way for attaining good quality work and also it avoids dispute at a later date. However, low rank in the contractor's response indicates the contractors view that the onus of providing a clear and unambiguous drawings and specifications on the owners and that will be provided in all cases without extra emphasis.

Similarly high ranks to equipping contractor's and subcontractors' men with proper tools, optimum resource estimation, and trained work force (rank 9, 11, and 14 respectively) in owner response indicate that owners attach high importance to these attributes in attaining desired quality performance, while their relatively low ranks (rank 36, 42, and 36 respectively) in contractor response may mean that contractors regard these activities as routine affairs and hence do not attach high importance to these activities.

For the seventh activity of the Table 7.6, 'coordinating hand over of work areas/service areas to other parties', difference between the mean values of owner and contractor responses could not be explained. In the relative term this activity

Table 7.6 Summary of ANOVA results of important activities between contractor and owner responses on Quality Criterion (Data arranged in the increasing mean value in owner response)

S. No.	Coordination activities	Mean of owner response	Rank of owner response	Mean of contractor response	Rank of contractor response	F-value	α
1.	Identifying or gathering information on defects, deficiencies, ambiguities, and conflicts in drawings and specifications and having them resolved	1.78	7	2.25	27	3.53	0.07
2.	Arranging for compliance with site instructions/directives from the engineer and revising programs/ordering material accordingly	1.82	8	2.84	55	14.27	0.00
3.	Equipping own men and subcontractors with tools, equipment, and resources	1.84	9	2.40	36	4.95	0.03
4.	Estimating the optimum resource requirements	1.88	11	2.48	42	7.63	0.00
5.	Arranging technical and behavioral training of human resources.	1.91	14	2.40	36	7.17	0.00
	Improving/altering/eliminating activities and considering better alternatives that may efficiently meet the project objectives.	1.94	19	2.46	40	3.77	0.06
7.	Coordinating hand over of work areas/service areas (such as plant rooms, service routes, etc.) to other parties	2.35	40	3.08	58	7.81	0.00

has received higher rank in owner responses than in contractor responses, but as such the rank of this activity is quite low in both owner and contractor responses.

7.5.4 No-Dispute Criterion

Important ANOVA results are summarized in Table 7.7, which lists only two activities. It can be commonly observed that contractors would like to have a better liaison between client and consultants to sort out any project related bottlenecks at the earliest. Similarly contractors would like to have a proper communication channel with outside authorities involved with the project. Contractors naturally view these as important to avoid dispute and hence keep these two activities at relatively higher rank (rank 10 and 34 respectively) when compared to owner

Table 7.7 Summary of ANOVA results of important activities between contractor and owner responses on No-dispute criterion (Data arranged in the increasing mean value in owner response)

S. No.	Coordination activities	Mean of owner response	Rank of owner response	Mean of contractor response	Rank of contractor response	F-value	α
1.	Acting as liaison with the client and the consultants	2.80	41	2.17	10	3.43	0.07
2.	Contacting outside authorities for testing, inspection and approval etc.	3.29	59	2.71	34	3.30	0.07

(rank 41 and 59). Owners might be thinking that since consultants and outside testing authorities are all ultimately paid by the owners' themselves, they will naturally have a tendency to protect the owners' interest.

7.5.5 Priority of Project Performance Criteria

Project success measurement depends to a great extent on the project performance criteria. We have discussed a number of performance criteria. These criteria can conflict with each other, which means there will often be trade-offs that must be agreed by all parties before the project is started (Wateridge 1998). In many projects there will be a large number of stakeholders, in which there is a need to identify which stakeholders are going to have the most influence in determining project success (Tuman 1986). From this, attention must be focused on important stakeholders if project success is to be accomplished.

It is thus important to set priority for the project performance criteria since for evaluating the project performance, the professionals may not give equal weightage to all the criteria. In order to get a generalized value of the weightages, Q9 of the first stage questionnaire (See Chap. 2) is framed. In the present study the priorities among five success criteria: schedule; cost; quality; no-dispute; and safety have been sought from the respondents. It is assumed that mean value of large responses to this question would give the relative preference in a generalized form for these performance criteria. From the scale structure of Q9, it can be observed that respondents are asked to give the rank 1 to the most important of these five criteria, rank 2 to the next most important and so on. These ranks are plotted on a cumulative frequency chart and are shown in Fig 7.1. The three highest ranked criteria: schedule, quality, and cost show dominance over the lowest two criteria: no-dispute, and safety.

This is also visible in Fig 7.2 which is showing the mean values of the responses obtained for Q9. As pointed earlier, the lower the mean value, the higher is its priority. From the mean values and the reversal of scale, the relative weights for these five criteria are found to be 0.268, 0.193, 0.242, 0.142, and 0.156 for schedule, cost, quality, no-dispute, and safety criteria respectively.

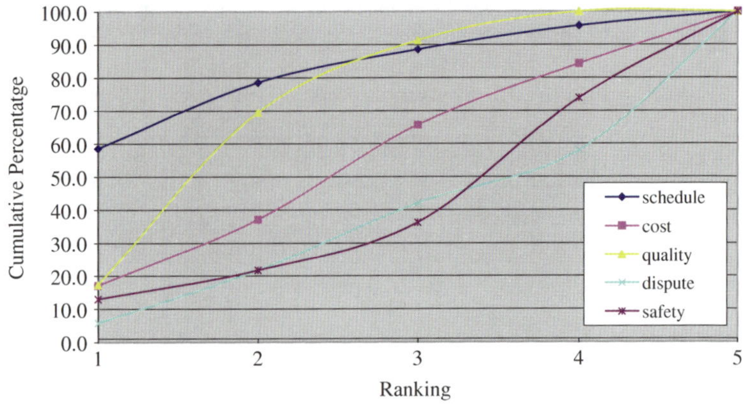

Fig. 7.1 Performance criteria cumulative frequency chart

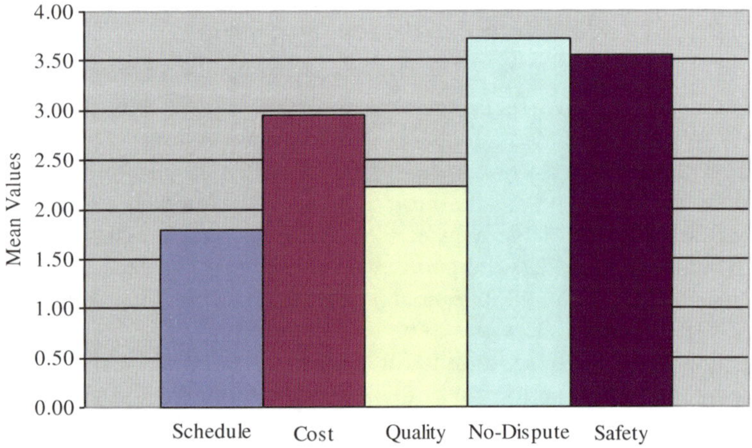

Fig. 7.2 Performance criteria ranking

When the safety criterion is omitted from the analysis, the revised relative weights attain a value of 0.317, 0.228, 0.287, and 0.168 respectively for schedule, cost, quality, and no-dispute criteria.

7.6 Identification of Important Coordination Activities for Overall Success

From the discussions presented in Sects. 7.5.1–7.5.4, it is seen that the important sets of coordinating activities vary with the performance criteria in question. In other words, certain coordination activities are found to be more important in a certain criterion. In order to evaluate the importance on an overall basis, the weights of different criterion obtained in the previous section were used. It was assumed that the score of contribution of any coordination activity is the weighted mean of the responses of the coordination activity received under the four performance criteria. The weighted mean so obtained for all coordination activities were arranged in an ascending order and are ranked starting from 1 to 59. From the scale structure used to collect data, it is clear that lower the mean value, the higher is its rank and the greater is its impact on influencing the performance criteria. Thus coordination activity with the lowest mean value is at rank 1 and so on. It may be recalled that the weights obtained for the schedule, cost, quality, and no-dispute criteria were 0.317, 0.228, 0.287, and 0.168 respectively.

The top ranking 20 coordination activities obtained on the basis of weighted mean were assumed to be requiring special attention for achieving coordination and thus these top 20 coordination activities (see Table 7.8) were selected for further study through the second stage questionnaire survey.

It is clear from the Table 7.8 that, 'implementing all contractual commitments' is the most important coordination activity perceived by Indian construction professionals. Indian contracts lay a large number of contractual requirements to

Table 7.8 Most important 20 coordination activities (Jha and Iyer 2006, with permission from Elsevier)

Id No.	Description of coordination activities
C_1	Implementing all contractual commitments
C_2	Arranging timely carrying out of all tests for inspections and approval by the engineer and maintaining records of the same
C_3	Arranging submission of samples of materials for approval by the engineer
C_4	Application of good technical practices
C_5	Preparation of a project quality plan in line with contract specification
C_6	Arranging remedial work methods and programs for executing in case of defect or damage
C_7	Identification of appropriate human resources, materials and equipments for the project
C_8	Estimation of the optimum resource requirements
C_9	Proper assignment of task to the available human resources for the project
C_{10}	Organization of resources (manpower, plant, and material) for effective utilization
C_{11}	Ensuring discipline among all employees
C_{12}	Resolving differences/conflicts/confusion among participants
C_{13}	Motivation of project participants
C_{14}	Development of a team spirit and receiving constructive input from all participants in the project
C_{15}	Identification of activities on critical path
C_{16}	Regular monitoring of critical path activities for adhering to schedule
C_{17}	Arrangement of required inputs like drawings, specifications, and technical details on time for execution
C_{18}	Agreement on detailed methods of construction with all the parties involved
C_{19}	Analysis of the project performances on time, cost and quality, and detecting variances
C_{20}	Monitoring the overall functioning of each section and department of the project

be fulfilled by the contractor executing a construction project. In view of this, the emergence of this activity as the most important activity is not surprising. Also, the gamut of this activity is so vast, it can easily encompass activities such as: 'arranging timely carrying out of all tests for inspections and approval by the engineer and maintaining records of the same', 'application of good technical practices', 'regular monitoring of critical path activities for adhering to schedule', and 'arrangement of required inputs like drawings, specifications, and technical details on time for execution' (ranked 2nd, 4th, 16th, and 17th in Table 7.8) within its fold.

The list of top coordination activities shown in Table 7.8 also point to resource management as important coordination activities. The effective coordination means proper resource management. This would consist of 'estimation of the optimum resource requirements' and assigning them in an appropriate manner (rank 8 and 9 respectively). Organizing the resources properly for their effective utilization is also an important coordination activity besides motivating project participants.

In addition to the above, some other important coordination activities are: identification of activities on critical path and monitoring them on a regular basis for the various project performance parameters such as time, cost and quality, and detecting variances.

In Fig 7.3, we show the essence of the top 20 coordination activities performed by a coordinator in a cyclic way which indicates that the majority of the activities are not a one-time affair but they are to be executed time and again for the entire duration of the project.

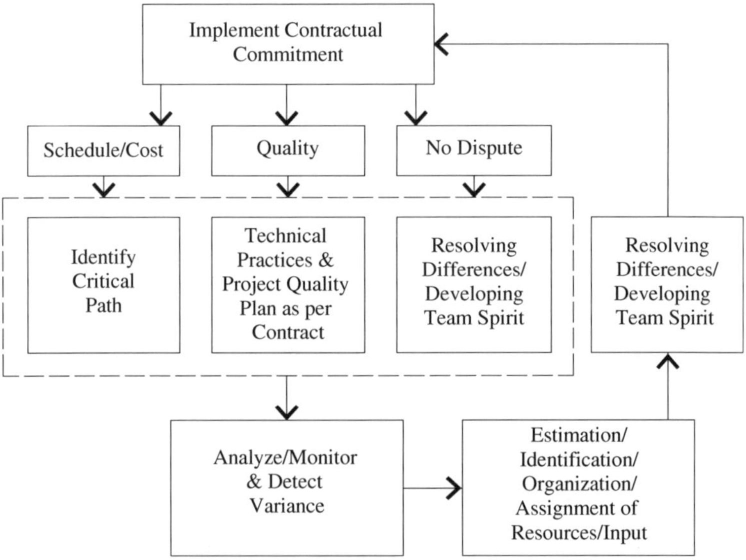

Fig. 7.3 Cyclic representation of coordination activities

7.7 Evaluation of Criticality of Coordination Activities

A second stage questionnaire survey was conducted to find the relative impact of top 20 coordination activities on project coordination. The top 20 coordination activities are considered as variables and the respondents were requested to evaluate the extent to which these activities were actually performed in the choice project on a 5-point scale. To remind the readers, the choice projects are the projects chosen by the respondents themselves and on which their responses are based. The details of the questionnaire were provided in Chap. 2. It may be recalled that in the 5-point scale, 1 represented 'unsatisfactorily done', 2 represented 'fairly done', 3 represented 'satisfactorily done', 4 represented 'nicely done', and 5 represented 'excellently done.' The respondents were also requested to specify the extent of contribution of good coordination (among project participants) on the outcome of the choice project. This response was also sought on the 5-point scale in which '1' indicated low contribution and '5' indicated high contribution.

Multinomial logistic regression analyses were conducted in which the coordination rating of the choice project was treated as dependent variable while the extent of coordination of the top 20 coordination activities (in terms of responses on these activities) were treated as independent variables. The relationship type of coordination rating with the coordination activities is as shown in Eq. 7.1.

$$\text{Coordination rating(C.R)} = f(C1, C2, C3, \ldots, C20) \quad (7.1)$$

The multinomial logistic regression was chosen because the responses obtained for both the dependent and independent variables were in discrete forms. It may also be noticed that the number of independent variables is large (20) and the multinomial logistic regression in such cases would require a large number of data sets of the order of 300–400 questionnaire responses as against the available 92 responses. Thus, the number of variables is required to be reduced and this needs a number of trials with different combination of fewer variables (Schwab 2003). Selection of variables using personal judgment is not acceptable in such studies. Thus in order to avoid personal biases in the selection of variables besides saving time in carrying out a large number of trials, hierarchical forward, and switching option of NCSS software was used. The results of multinomial logistic regression are presented in Table 7.9.

Table 7.9 clearly shows that depending on the level of coordination ratings, coordination activities vary. For example, coordination activities C_8 (Estimation of the optimum resource requirements) and C_{18} (Agreement on detailed methods of construction with all the parties involved) are significant at coordination rating $CR = 1$, while, C_6 (Arranging remedial work methods and programs for executing in case of defect or damage) and C_8 (Estimation of the optimum resource requirements) are significant at $CR = 2$. Only one coordination activity C_5 (Preparation of a project quality plan in line with contract specification) has emerged to be significant at $CR = 3$ while five coordination activities: C_5

Table 7.9 Summary of important results of multinomial logistic regression between important coordination activities and coordination rating (Jha and Iyer 2006, reproduced with permission from Elsevier)

Coordination rating	Variable	Log of odds ratio, B	Std. error, SE	Wald stat = $(B/SE)^2$	Sig. level, α	Odds ratio, e^B	p	Q	Δp	Δq
(1)	(2)	(3)	(4)	(5)	(6)	(7)	(8)	(9)	(10)	(11)
1.00	Intercept	15.622	4.716	10.972	0.001	–	–	–	–	–
	C8	−1.347	0.722	3.485	0.062	0.260	0.21	0.79	−0.14	0.14
	C18	−3.104	1.298	5.716	0.017	0.044	0.04	0.96	−0.04	0.04
2.00	Intercept	12.704	3.923	10.489	0.001	–	–	–	–	–
	C6	−2.768	0.895	9.571	0.002	0.062	0.06	0.94	−0.05	0.05
	C8	−1.216	0.682	3.179	0.075	0.296	0.23	0.77	−0.15	0.15
3.00	Intercept	9.713	3.310	8.612	0.003	–	–	–	–	–
	C5	−1.393	0.602	5.351	0.021	0.248	0.20	0.80	−0.14	0.14
4.00	Intercept	5.194	3.009	2.980	0.084	–	–	–	–	–
	C5	−1.302	0.570	5.228	0.022	0.272	0.21	0.79	−0.14	0.14
	C8	−1.086	0.485	5.012	0.025	0.337	0.25	0.75	−0.15	0.15
	C14	1.882	0.669	7.917	0.005	6.570	0.87	0.13	0.11	−0.11
	C17	1.262	0.550	5.260	0.022	3.531	0.78	0.22	0.15	−0.15
	C18	−1.371	0.694	3.900	0.048	0.254	0.20	0.80	−0.14	0.14

(Preparation of a project quality plan in line with contract specification), C_8 (Estimation of the optimum resource requirements), C_{14} (Development of a team spirit and receiving constructive input from all participants in the project), C_{17} (Arrangement of required inputs like drawings, specifications, and technical details on time for execution), and C_{18} (Agreement on detailed methods of construction with all the parties involved) have emerged to be significant at CR = 4. Thus out of 20 coordination activities, six activities C_5, C_6, C_8, C_{14}, C_{17}, and C_{18} have emerged to be significant at different coordination ratings.

The Chi square value of 83.004 at 24 degrees of freedom is found to be highly significant (significance level = 0.000),which means that the null hypothesis that all effects of the independent variable are zero can be rejected (Whitehead 1998). The Nagelkerke R^2 value obtained from the analysis is 0.684 which indicates that the multinomial logistic model in this case performs well for the coordination rating range from 1 to 4 with rating 5 being the reference category unlike the project performance that were measured in 10 point scale. The percent correct prediction is 65 % in this case.

At CR = 1, the B-values for C_8 and C_{18} are −1.347 and −3.104 respectively. The negative values indicate that any increase in the variable values of C_8 and C_{18} would significantly decrease the likelihood of project coordination rating being at level 1 and thus in turn chances of achieving a coordination rating of 5 increases. In the present case the negative value of B corresponding to coordination rating of 1 in Col 3 of Table 7.9, indicates that any increase in variable value of C_8 (Estimation of the optimum resource requirements); and C_{18} (Agreement on detailed methods of construction with all the parties) would significantly decrease the likelihood of project coordination rating being at the level 1. The values of Δp and Δq in Col 10 and Col 11 indicate that with one unit rise in the value of C_8, and C_{18}, the probability of performance to remain at level 5 decreases by 14 and 4 % respectively or conversely the probability of achieving the coordination rating 5 (improving the coordination) will increase by 14 and 4 %. Similarly, at CR = 2, the values of Δp and Δq in Col 10 and Col 11 indicate 5 and 15 % increase in probability values of bettering of coordination rating with unit increase in the variable values of C_6 (Arranging remedial work methods and programs for executing in case of defect or damage); and C_8 (Estimation of the optimum resource requirements) respectively. At CR = 3, the Δp and Δq values indicate that with one unit increase in the value of C_5 (Preparation of a project quality plan in line with contract specification), the probability of bettering the coordination rating increases by 14 %.

At CR = 4, unit increase in the values of three variables: C_5 (Preparation of a project quality plan in line with contract specification); C_8 (Estimation of the optimum resource requirements) respectively; and C_{18} (Agreement on detailed methods of construction with all the parties) enhance the probability of bettering the coordination rating. The remaining two variables: C_{14} (Development of a team spirit and receiving constructive input from all participants in the project); and C_{17} (Arrangement of required inputs like drawings, specifications, and technical details on time for execution), increase the probability to retain the coordination level at the existing coordination rating of 4.

The analysis results in the Table 7.9 also indicate that two most important coordination activities are C_8 and C_{18}, which show their potential of continual improvement of coordination levels if handled properly toward enhancement.

7.8 Assessment of Probabilities of Coordination Ratings

Application of multinomial logistic regression models in the previous section and in other areas of application of this model (Schwab 2003, NCSS) reveal that if the levels of achievement of the six coordination activities (C_5, C_6, C_8, C_{14}, C_{17} and C_{18}) are given, the probabilities of achieving a particular level of coordination in a project can be estimated as per the equations given below.

$$\Pr ob\,(CR = 1) = Y_1 = \frac{1}{1 + e^{-(15.62 - 1.22 \times A_{C5} - 0.40 \times A_{C6} - 1.34 \times AC_8 - 0.49 \times A_{C14} + 1.48 \times A_{C17} - 3.10 \times A_{C18})}} \tag{7.2}$$

$$\Pr ob\,(CR = 2) = Y_2 = \frac{1}{1 + e^{-(12.70 - 0.38 \times A_{C5} - 2.76 \times A_{C6} - 1.21 \times A_{C8} - .019 \times A_{C14} - 0.82 \times A_{C17} + 0.61 \times A_{C18})}} \tag{7.3}$$

$$\Pr ob\,(CR = 3) = Y_3 = \frac{1}{1 + e^{-(9.71 - 1.39 \times A_{C5} - 0.49 \times A_{C6} - 0.28 \times A_{C8} - 0.18 \times A_{C14} + 0.09 \times A_{C17} - 0.29 \times A_{C18})}} \tag{7.4}$$

$$\Pr ob\,(CR = 4) = Y_4 = \frac{1}{1 + e^{-(5.19 - 1.30 \times A_{C5} - 0.57 \times A_{C6} - 1.08 \times A_{C8} 1.88 \times A_{C14} + 1.26 \times A_{C17} - 1.37 \times A_{C18})}} \tag{7.5}$$

where

- Y_1, Y_2, Y_3, and Y_4 denote the probabilities of achieving coordination of a particular rating 1, 2, 3 and 4, with rating 5 being of reference category;
- CR = Coordination Rating, 1, 2, 3, or 4, and
- A_{C5}, A_{C6}, A_{C8}, A_{C14}, A_{C17} and A_{C18} are given value of achievement of coordination activities C_5, C_6, C_8, C_{14}, C_{17}, and C_{18} as given in Q8 of second stage questionnaire. These values for a given project can be obtained by seeking responses on these variables on a five point scale as given in the Q8 from the project participants and averaging them or from any performance auditor of the project.

The above equations suggest that for any value of the set of the six variables (C_5, C_6, C_8, C_{14}, C_{17} and C_{18}) will give four probabilities, each describing chances of occurrence of coordination rating level 1, 2, 3, 4, or 5. Since the six variables too have five levels of achievements (1, 2, 3, 4, and 5), there can be 6^5 (=7,776) combinations of values, and each set of values leads to four probabilities as stated earlier, understanding all of them may not be possible. Therefore, application of the model is illustrated through a simple example, where A_{C5}, A_{C6}, A_{C8}, A_{C14}, A_{C17}, and A_{C18} are assumed to have values, either 1 for all, or 2 for all, and so on. For example, if the variable values A_{C5}, A_{C6}, A_{C8}, A_{C14}, A_{C17} and A_{C18} are all low, say 1 and it is desired to estimate the probability of achieving coordination rating 2, Eq. 5.3 can be used for this purpose. The corresponding probability to achieve coordination rating 2 from the equation is estimated as 0.999 (close to 1). When the variable values are changed to 2, the probability to achieve coordination rating of 2 becomes 0.26. Conversely the probability to achieve the coordination rating 5 becomes 0.74 (1−0.26). Similarly other probability values can also be calculated by keeping uniform values of 3, 4, or 5 to the variables kept. These probabilities can be better visualized by the representative probability curve plotted for this case in Fig 7.4. The probabilities values obtained also infer that with low level of achievements of coordination activities, there will be greater

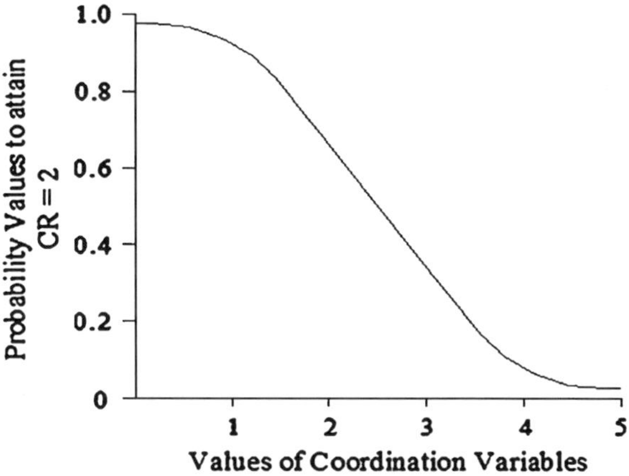

Fig. 7.4 Probability curve at different values of coordination variables

chances of rating of overall coordination being at low level and by improving some of the critical activities listed above; there are chances of coordination level bettering from the existing status.

7.9 Taxonomy of Coordination Activities

In order to understand better the several coordination activities by a few common properties, Saram and Ahmed (2001) have tried to identify them in groups and expressed the difficulty faced by them. They have employed a number of combinations based on their intellectual wisdom to have a meaningful classification of groups. However, there are chances of personal bias in such combination. To avoid this, the present study adopts factor analysis. The utility of factor analysis in such situations has already been discussed. This technique is employed to analyze the responses on the 20 coordination activities of second level questionnaire. Since the factors extracted initially are orthogonal and not amenable to any interpretations, varimax rotation is performed. A total of four factors are extracted. Associated variables in different factors and their factor loadings are summarized in Table 7.10. The common characteristics of variables of different factors that give names to the factors are discussed below.

7.9.1 Factor_1 Planning

On close scrutiny of activities emerging in Factor_1 in Table 7.10 it is observed that all the activities are predominantly planning related activities and generally the planning section or planning department staff of the organization carries out these actions. However, planning section or department will also be headed by the

Table 7.10 Factor profile of construction coordination activities (Jha and Misra 2007, with permission from Taylor and Francis)

	Factor loading
Factor_1 Planning (Variance explained 19.08 %)	
Monitoring the overall functioning of each section and department of the project	0.82
Analysis of the project performances on time, cost and quality, and detecting variances from the schedule/requirements and dealing with their effects considering time and resource constraints	0.71
Arrangement of required inputs like drawings, specifications, and technical details on time for execution	0.64
Agreeing on detailed methods of construction with all the parties involved	0.60
Application of good technical practices	0.51
Preparation of a project quality plan in line with contract specification	0.50
Arranging remedial work methods and programs for executing in case of defect or damage	0.47
Implementation of all contractual commitments	0.44
Identification of appropriate human resources, materials, and equipments for the project	0.41
Factor_2 Resource Handling (Variance explained 18.71 %)	
Estimation of the optimum resource requirements	0.86
Identification of appropriate human resources, materials, and equipments for the project	0.73
Development of a team spirit and receiving constructive input from all participants in the project	0.65
Proper assignment of task to the available human resources for the project	0.61
Organization of resources (manpower, plant, and material) for effective utilization	0.55
Motivation of project participants	0.53
Identification of activities on critical path	0.45
Application of good technical practices	0.45
Preparation of a project quality plan in line with contract specification	0.44
Factor_3 Contract Implementation (Variance explained 17.41 %)	
Arranging submission of samples of materials for approval by the engineer	0.84
Arrangement for timely carrying out of all tests for inspections and approval by the engineer and maintaining records of the same	0.81
Implementation of all contractual commitments	0.61
Preparation of a project quality plan in line with contract specification	0.53
Identification of activities on critical path	0.52
Regular monitoring of critical path activities for adhering to schedule	0.48
Application of good technical practices	0.47
Factor_4 Team Building (Variance explained 13.32 %)	
Resolving differences/conflicts/confusion among participants	0.81
Ensuring discipline among all employees	0.67
Motivation of project participants	0.57
Organization of resources (manpower, plant, and material) for effective utilization	0.48

Project Manager, who is apparently responsible for overall steering of the project to his liking. This factor alone explains a variance of about 19.08 % in the total variance of 68.53 % explained by the factor analysis. In the relative term, it can be said that this factor accounts for 27.84 % (=19.08/68.53) among all factors. It means that PLANNING holds the key for achieving proper coordination and if handled carefully will fetch great amount of success.

7.9.2 Factor_2 Resource Handling

The next important factor that emerges is resource handling. The activities under this factor predominantly are resource related e.g. estimation, identification, organization, and proper assignment of resources. It also has activities such as motivating the human resources, and developing the team spirit among themselves, so crucial for attaining the desired performance level. Proper resource handling will automatically take care of the last three activities appearing under this factor. This factor alone explains a variance of 18.71 % in absolute terms and 27.53 % in relative **terms.**

7.9.3 Factor_3 Contract Implementation

This factor contains the activities that are supposed to be done by the executing agencies under contract terms and conditions. These activities are timely submission of samples for approval, timely carrying out of all tests, implementing all contractual commitment, preparation of quality plan, and applying good technical practices. Identification and monitoring of critical path activities are part and parcel to achieve schedule completion, which is one of the major contract commitments as time is invariably the essence of modern contract. This factor explains a variance of 17.41 % in absolute scale and 25.40 % in relative scale.

7.9.4 Factor_4 Team Building

The last factor is Team Building as explained by most activities falling under Factor_4 in Table 7.10. Building team is an essential part for the execution of modern day project having multiple participants. During the progress of the work differences are bound to crop up among different participants due to clashes in their interests. In these situations, resolving of the differences/conflict, and motivation of participants come into picture. Ensuring discipline among the participants and organizing the resources properly according to the expertise of the resources are all aimed to enhance greater teamwork. This factor explains 13.32 % of variance and holds relative responsibility of 19.44 %.

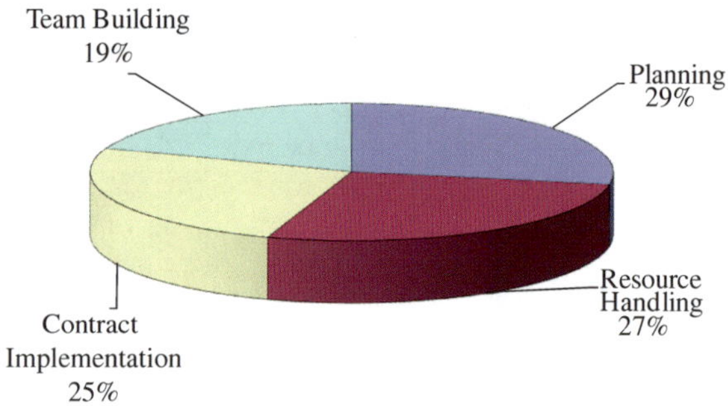

Fig. 7.5 Pie chart showing contribution of factors in coordination activities

It can be inferred from the factor extraction and variances explained that the project coordination is not an isolated and independent activity, but is a typical management function having its inherent role of varying degrees in all the major management activities that are broadly represented by the above four factors, i.e., Planning, Resource handling, Contract implementation, and Team building. The variances explained by these factors in relative terms are presented in the pictorial form as a pie chart in Fig 7.5.

It is observed from the literature in mathematics, the variances explained by the factors are good measure to explain the importance, however a few feel that in factor analysis like variable are dumped together based on the response pattern and it would not represent the order of criticality of the factors (Iyer 1996). Hence to understand the criticality of the above four factors, a stepwise multiple regression was performed between the responses on coordination among project participants of the choice project as the dependent variable and the above four factors as the independent variables. Factor scores for each respondent formed the data set for four factors and the stepwise regression analysis was carried out in SPSS software. Details of regression results are shown in Table 7.11. This table shows the unstandardized coefficient B, standardized regression coefficients β, significant level α, the value of R^2, and the change in R^2 value (ΔR^2).

It can be observed from the Table 7.11 that all the factors have emerged to be significant in the analysis. The criticality of the factor is judged by its standardized coefficient β value and in this respect, the most critical factor to achieve good coordination among project participants has emerged to be 'Resource Handling', and this is followed by 'Planning', 'Team Building', and 'Contract Implementation'.

Table 7.11 Summary of stepwise regression results (Jha and Misra 2007, with permission from Taylor and Francis)

Independent variables	B	β	α	R^2	ΔR^2
Factor_1: Planning	0.348	0.287	0.016	0.082	–
Factor_2: Resource handling	0.463	0.382	0.001	0.228	0.146
Factor_3:Contract implementation	0.226	0.187	0.082	0.263	0.035
Factor_4: Team building	0.273	0.225	0.032	0.314	0.051
Constant	3.671				

7.10 Summary and Conclusions

A number of activities need to be carried out in order to achieve day-to-day coordination in a construction project. In this study 59 such activities have been identified and they have been evaluated for their degree of importance on the four performance criteria: *schedule*, *cost*, *quality*, and *no-dispute*. The important findings of this chapter are given below.

- The analysis result in this chapter has shown that relative importance of coordination activities vary depending on the performance measure employed for measuring the success. For example, when schedule criteria is given the supreme priority, the most important activity for a coordinator is 'regular monitoring of critical path activities for adhering to schedule' whereas when cost is taken as the prime performance criteria 'monitoring the budget on all activities and taking corrective action' takes supreme importance. Similarly 'application of good technical practices' and 'implementation of all contractual commitments' turn out to be the most important coordination activities for quality and no-dispute performance criteria respectively.
- Since all 59 coordination activities are not of high/equal importance 20 important coordination activities have also been identified for achieving day-to-day coordination at project sites. Application of multinomial logistic regression of the responses on these 20 activities with actual level of coordination achieved for a project identifies that six coordination activities are significantly important in influencing the coordination rating of a project. These activities are: C_5—preparation of a project quality plan in line with contract specification; C_6—arranging remedial work methods and programs for executing in case of defect or damage; C_8—estimation of the optimum resource requirements; C_{14}—development of a team spirit and receiving constructive input from all participants in the project; C_{17}—arrangement of required inputs like drawings, specifications, and technical details on time for execution; and C_{18}—agreement on detailed methods of construction with all the parties. Most important of them are C_8 and C_{18}, which show their potential of continual improvement of coordination levels if handled properly toward enhancement.
- Project coordination is not an isolated and independent activity, but is a typical management function having its inherent role of varying degrees in all the major management activities that are broadly represented by the four factors, i.e., Planning, Resource handling, Contract implementation, and Team building. The most critical factor explaining the highest variance in achieving coordination turns out to be 'Resource handling'.

However, effort needed to achieve proper coordination at sites is dependent on a number of elements. Also achieving proper coordination demands specialized skill in the person coordinating the projects. These topics have been taken up in the next chapter.

References

Grigg, N. S. (1993). New paradigm for coordination in water industry. *Journal of Water Resources Planning and Management, ASCE, 119*(5), 572–587.

Iyer, K. C. (1996). Identification and evaluation of dispute-prone clauses in Indian construction contracts. A Doctoral Thesis submitted at Indian Institute of Technology, Madras.

Jha, K. N.,& Iyer, K. C. (2006). Critical determinants of project coordination, *International Journal of Project Management*, *24*(4), 314–322.

Jha, K. N., & Misra, S. (2007).Ranking and classification of construction coordination activities in Indian projects, *Construction Management and Economics*, Taylor and Francis, *25*(4), 409–421.

Lam, E. L. (1991). Hardwiring coordination among land use and transportation agencies. *Journal of Urban Planning and Development, ASCE, 117*(4), 168–180.

Lammie, J. L., & Shah, D.P. (1980). Project management—Pulling it all together. *Journal of Transportation Engineering Division*, *ASCE, 116* (TE4), 437–451.

Low cost (1993). *India Today*, May 31, *XVIII*(10), 95.

Nam, C. H., & Tatum, C. B. (1992). Non contractual methods of integration on construction projects. *Journal of Construction Engineering and Management, ASCE, 118*(2), 385–398.

Pinto, J. K., & Slevin, D. P. (1988). Project success: definitions and measurement techniques. *Project Management Journal, 19*(1), 67–72.

Rad, P. F. (1979). Delays in construction of nuclear power plant. *Journal of the Energy Division, Proc of ASCE, 105*(EY/1), 33–46.

Ruchelman, L. (1980). Coordinating tall building development. *Journal of the Urban Planning and Development Division*, *ASCE, 106*(UP1), 89–101.

Saram, D. D. D., & Ahmed, S. M. (2001). Construction coordination activities: What is important and what consumes time? *Journal of Management in Engineering, ASCE, 17*(4), 202–213.

Schwab, A. J. (2003). http://www.utexas.edu/courses/schwab

Sheeran, F. B. (1976). *Management essentials for public works administrators*. Chicago: American Public Works Association.

Tuman, J. (1986). Success modeling: A technique for building a winning project team. In *1986 Proceedings of the Project Management Institute*: Montreal, *Project Management Institute*. Drexel Hill, (pp. 94–108).

Wang, Y. (2000). Coordination issues in Chinese large building projects. *Journal of Management in Engineering, ASCE, 16*(6), 54–61.

Wateridge, J. (1998). How can IS/IT projects be measured for success? *International Journal of Project Management, 16*(1), 59–63.

Whitehead, J. (1998). http://www.personal.ecu.edu/whiteheadj

Chapter 8
Other Issues in Project Coordination and Traits of a Project Coordinator

Abstract Elements affecting project coordination have been identified and evaluated based on their criticality and the increase in coordination effort due to these elements. Literature on traits of project manager is reviewed and distinction between the role of project manager and project coordinator is made. There are very few studies which lay down the required traits of a project coordinator. Common traits identified from literature are evaluated based on their relative importance for a project coordinator through the analysis of questionnaire responses. The important traits have converged into three major skill categories: team building skill, contract implementation skill, and project organization skill.

8.1 Introduction

Coordination is required in almost every stage of a project and considerable time is consumed in coordination. Prudent project managers realize that time spent on coordination is an investment, which bears fruits through success of projects. It was seen in the previous chapter that coordination by itself is not an independent activity, but it is associated with some main activities of the project. Hence while calculating the time and effort spent on any activity, coordination is considered to be an integral part of the main activity rather than being measured separately. It is probably this reason that although coordination is a very important function of management, it has not received adequate attention in the available literature. In order to understand the aspects of coordination in detail, key activities for achieving coordination in a construction project have been identified in the previous chapter and their importance have been evaluated on the four performance

This chapter is based on 'What attributes should a project coordinator possess?'Construction Management and Economics, 24(9), 2006, with permission from Taylor and Francis.

K. N. Jha, *Determinants of Construction Project Success in India*, Topics in Safety, Risk, Reliability, and Quality 23, DOI: 10.1007/978-94-007-6256-5_8, measuring criteria—*schedule; cost; quality*; and *no-dispute*. These activities have been also recognized by some latent common properties called as factors that would help project managers to understand the role of latent properties. To explore further on other issues related coordination the following objectives have been set:

- identifying key elements affecting coordination and to evaluate the effect of these elements on the coordination effort
- identifying and analyzing dominant skills/traits of a project coordinator for success of a project.

8.2 Identification of Key Elements Affecting Coordination

From the literature and personal interviews with professional 23 key elements affecting coordination among project participants are identified (Iyer and Jha 2004). These elements generally relate to project specific, contractor and subcontractor specific, tender specific, client specific, labor specific, finance specific, material, and equipment specific issues. These key elements are then taken in the first stage questionnaire (Q12 as explained in Chap. 2) where the respondents are asked to give their opinions on these elements on a five point semantic scale on two qualitative parameters: (a) their intrinsic *characteristics* and (b) their *influence on coordination effort*. *Intrinsic characteristics* of any element is defined by the qualitative parameters such as 'Highly critical'; 'Critical'; 'Important but not critical'; 'Not so important'; or 'Irrelevant' represented by 1, 2, 3, 4, and 5 respectively.

A close examination of all coordination affecting elements reveal that not all the elements would cause an increase in the coordination effort, but some of them are favorable decreasing the coordination efforts in a project. To avoid any personal bias by intellectual guess, responses on the 23 key elements are sought through the second part of the question, *influence on coordination effort* (Q12 in Appendix A). The responses are sought on the five-point scale in which 1, 2, 3, 4, and 5 represent 'significant increase', 'marginal increase', 'significant decrease', 'marginal decrease', and 'inconsequential' respectively. However, in order to have consistency with responses of other variables of the questionnaire, scale of this question is altered for analysis purposes. The scale after alteration would read '1' as 'significant increase', '2' as 'marginal increase', '3' as 'inconsequential', '4' as 'marginal decrease', and '5' as 'significant decrease'.

As indicated in Chap. 3, a total of 114 responses are received out of 400 questionnaires sent. Mean values of responses on the 23 elements are calculated separately for *intrinsic characteristic* and *influence on coordination effort* and these elements are arranged in increasing order of the mean values and ranked separately for *intrinsic characteristics* and *influence on coordination effect*.

For interpretation purpose the elements having mean values significantly less than or equal to 1.5 ($\mu \leq 1.5$) in the *intrinsic characteristic* are considered to be highly critical, while those with mean values greater than 1.5 but less than or equal to 2.5 ($1.5 < \mu \leq 2.5$) are called as critical and all other elements having mean values greater than 2.5 but less than or equal to 3.5 ($2.5 < \mu \leq 3.5$) as important.

Similarly in the *influence on coordination efforts* category, those elements having mean values significantly less than or equal to 1.5 ($\mu \leq 1.5$) are considered to cause significant increase in coordination efforts in performing the task, while those having mean values greater than 1.5 but less than or equal to 2.5 ($1.5 < \mu \leq 2.5$) are considered to cause marginal increase in coordination efforts in performing the task and the remaining elements having mean values greater than 2.5 but less than or equal to 3.5 ($2.5 < \mu \leq 3.5$) are the ones which are called inconsequential, for they neither increase nor cause any saving in the coordination efforts. The mean values of responses on key elements affecting coordination ranged from 1.20 to 3.12 in *intrinsic characteristic* and 1.63 to 3.29 in *influence on coordination efforts*. To identify the criticality of elements, the 23 elements are sorted out based on their mean values. Table 8.1 summarizes all elements in the increasing order of mean values under two parameters.

It can be observed from Col 2 and Col 3 of the Table 8.1 that there are three elements in the highly critical category (with $\mu \leq 1.5$)and 19 elements($1.5 < \mu \leq 2.5$)fall in critical category group leaving only one element in noncritical but important category. Similarly Col 4 and Col 5 indicate that three elements having $\mu \leq 1.5$ cause significant increase in coordination effort while 16 elements ($1.5 < \mu \leq 2.5$) falling in the category of elements that require marginal increase in coordination efforts leaving four elements having mean values between 2.5 and 3.5 ($2.5 < \mu \leq 3.5$) indicating neither increase nor saving in coordination effort. It is interesting to note that two out of three highly critical elements are the ones that require significant increase in the coordination efforts. Hence, in order to achieve success in a project, the person responsible for coordinating needs to put very high attention in these two highly critical elements :*nonavailability of finance for the project*; and *very high degree of hazard associated with the project.*

Identification of the above two elements does not allow to relax the attention required for other elements as excepting two elements all elements are found critical and almost all require extra coordination efforts of varying degrees. In the next section the other issue of coordination, the traits of the project coordinator is discussed.

8.3 Review of Traits of a Project Manager

A project manager is a person with specific accountability and is formally appointed to manage a project for achieving defined project objectives within allocated resources. A project manager has access to, and a formally defined relationship with, the project leader to which the specific project has been assigned (www.tbs-sct.gc.ca). Project managers play a very important role in success of a project and recognizing this many studies have been conducted to find their required traits.

Table 8.1 Summary of mean responses on key coordination elements and their categorization (**a**) Variables with $\mu \leq 1.5$ (Highly critical category), and (**b**) Variables with $1.5 < \mu \leq 2.5$ (Critical category) (**c**) Variables with $\mu > 2.5$ (Important category)

S. No. (1)	Intrinsic characteristics (2)	Mean (3)	S. No. (4)	Influence on coordination effort (5)	Mean (6)
(**a**) *Variables with $\mu \leq 1.5$ (Highly critical category)*			*Variables with $\mu \leq 1.5$ (significant increase in coordination efforts)*		
1.	Nonavailability of finance for the project	1.20	1.	Increasing project size (in terms of contract price)	1.63
2.	Very high degree of hazard associated with the project	1.49	2.	Very high degree of hazard associated with the project	1.72
3.	Lack of PM's (Project Manager) experience in handling the project/work	1.66	3.	Nonavailability of finance for the project	1.75
(**b**) *Variables with $1.5 < \mu \leq 2.5$ (Critical category)*					
4.	Inadequate period of completion given in the contract	1.73	4.	Increasing project complexity (e.g. complex architectural features)	1.78
5.	Inadequate drawings and details	1.79	5.	Nature of project from regular job to a more complex job	1.78
6.	Nature of project from regular job to a more complex job	1.90	6.	Inadequate period of completion given in the contract	1.83
7.	Delay due to involvement of many agencies/statuary bodies for approval	1.90	7.	Delay due to involvement of many agencies/statuary bodies for approval	1.90
8.	Increasing project complexity (e.g. complex architectural features)	1.93	8.	Increasing complexity of business sector from conventional to more advanced business sector	1.93
9.	Increasing project duration	1.97	9.	Inadequate drawings and details	1.97
10.	Supply of resources like material and equipment in the client' scope	1.98	10.	Lack of PM's (Project Manager) experience in handling the project/work	2.03
11.	Contract type—from conventional one (such as item rate, cost plus, lump sum) to complex ones (turnkey, BOT etc.)	2.12	11.	Increasing project duration	2.29
12.	Increasing complexity of business sector from conventional to more advanced business sector	2.12	12.	Contract type—from conventional one (such as item rate, cost plus, lump sum) to complex ones (turnkey, BOT etc.)	2.32
13.	Increasing project size (in terms of contract price)	2.14	13.	Presence of labor union	2.32
14.	Presence of reliable subcontractor	2.17	14.	Supply of resources like material and equipment in the client' scope	2.33

(continued)

Table 8.1 (continued)

S. No. (1)	Intrinsic characteristics (2)	Mean (3)	S. No. (4)	Influence on coordination effort (5)	Mean (6)
15.	Selection of knowledgeable and motivated subcontractors	2.17	15.	Increasing restriction in working hours	2.35
16.	Presence of labor union	2.27	16.	The location of the project from the native states to other states/countries	2.42
17.	Increasing restriction in working hours	2.28	17.	Requirement of stage passing	2.71
18.	Presence of liquidated damage clause in contract	2.40	18.	Presence of liquidated damage clause in contract	2.73
19.	Requirement of stage passing	2.44	19.	The type of client from Government bodies to Private bodies	2.75
20.	Previous experience with the associated agencies	2.60	*Variables with $2.5 < \mu \leq 3.5$ (Inconsequential, i.e., neither increase nor saving in coordination efforts)*		
21.	The location of the project from the native states to other states/countries	2.63	20.	Previous experience with the associated agencies	3.00
22.	The type of client-from Government bodies to Private bodies	2.66	21.	Selection of knowledgeable and motivated subcontractors	3.05
(c) *Variables with $\mu > 2.5$ (Important category)*			22.	Presence of reliable subcontractor	3.09
23.	Absence of liquidated damage clause	3.12	23.	Absence of liquidated damage clause in contract	3.29

Although, the terms project manager, project coordinator, construction manager, project administrator, and project controller are used quite interchangeably and all of them appear to have very similar kind of role, yet the intensity of their job requirement and expectations from them vary (Kerzner 2002). For example, the typical responsibilities of a project manager and a project coordinator both include: coordinating and integrating of subsystem tasks; assisting in determining technical and manpower requirements, schedules and budgets; and measuring and analyzing project performance regarding technical progress, schedules, and budgets. However a project manager is supposed to play a stronger role in project planning and controlling. While a project manager is also responsible for negotiating; developing bid proposal; establishing project organization and staffing; and providing overall leadership to the project team in addition to profit generation and new business development, a project coordinator is seldom entrusted with these responsibilities. In fact, the project coordinator's role is to augment the project managers' visibility for larger projects (Forsberg et al. 1996). A project coordinator is chartered as a representative of the project manager who proactively ensures future events will occur as planned. They signal problem areas and recommend solutions. According to Forsberg et al. (1996), Project coordinator's function is to

- Know how the organization "works"
- Provide expediting help to the project and support organizations
- Provide independent assessment of project information and status to the project manager
- Ensure planning and milestones are satisfied
- Ensure control procedures are being adhered to.

According to Katz and Kahn (1978), an effective project manager should possess essentially three skills: technical skills, human relationships skills, and conceptual skills. While technical skills include the ability to apply knowledge in a given field, such as engineering and finance and so on, human relationships skills involve the ability to communicate efficiently and to maintain a harmonious working group. The ability to motivate employees falls into human relationships skills category. Finally, conceptual skills include the ability to perceive the project as a system by keeping a global perspective and not thinking of only one aspect at a time.

The above model has led to a number of debates on the extent to which a project manager needs technical skill. While it is understandable to have a technical expert as a project manager in case of a small project that involves knowledge of only one small specialist area, for larger projects involving multiple disciplines searching for a technical expert may not be a wise option (Goodwin 1993). This is not to say that technical skills are not needed at all in larger projects but the emphasis should be more on managerial skills of a project manager. Technical skills in larger projects are needed to appreciate the full implications of the project, which a project manager obtains as expert advice on as and when basis. Some researchers are also of the opinion that project manager should not be a technical expert due to the apprehension that project manager would engage himself in too much technical details and may not be able to do justice to other aspects of the project (Katz 1974; Goodwin 1993; Kerzner 2002).

El-Sabaa (2001) has analyzed the relative importance of the three skill groups advocated by Katz and Kahn. Human skill with a percentile score of 85.3 has emerged to be the most essential project manager skill. Conceptual and organizational skill with a percentile score of 79.6 represents the second essential project manager skill. Technical skill with a percentile score of 50.46 has emerged the least essential project manager's skill relatively.

Odusami (2002) concludes through the analysis of a questionnaire survey conducted among the clients, consultants, and contractors that for a client the most important skill of an effective project leader is decision-making; for a consultant the most important skill is leadership and motivation, and for the contractors, communication is the most important skill. Laufer et al. (1999) opine that a project manager's principle role is to manage his/her team's decision-making and not to make his own decisions.

Various researchers have stressed the need for different types of skills required by a project manager in order to make the project successful. Their findings are either based on their experiences or based on empirical researches. Amongst the first category, we have the skills suggested by Stuckenbruck (1976), Kerzner (2002), Fryer (1979), Adams and Barndt (1978). The details of these suggested skills are presented in Table 8.2.

Among the various literature cited above the empirical research has been by Spitz (1982). She concluded that the priority of skills of a project manager vary depending on the phase in which the project presently exists. She has also tried to assess the skills needed in each phase of a project.

Tarricone (1992) believes that although some are born leaders, most managers are actually created. Some of the ways in which a future project manager can be groomed start right from undergraduate training; graduate and continuing education; employer in-house training programs; and by seat-of-the-pants experience.

8.4 Important Traits of a Project Coordinator

As can be observed from the previous section, a number of studies have been undertaken to address the skill requirement of a project manager, however very few studies have been undertaken to explore the skill requirement of a person coordinating a construction project. Though a project manager is also responsible for project coordination to an extent, in a large project, project manager needs assistance from project coordinator who takes care of coordination aspect of the project. The need for a project coordinator has also been recognized in the past and authors such as Forsberg et al. (1996) and Kerzner (2002) have tried to distinguish the roles and responsibilities between a project manager and a project coordinator.

To illustrate, let's consider a multi-disciplinary large project which requires involvement of personnel from different departments like civil, electrical,

Table 8.2 Summary of project manager's skill identified in different studies (Jha and Iyer 2006, with permission from Taylor and Francis)

Authors	Skill description
Gaddis (1959)	Project manager needs solid basic experience in the relevant field and should be a leader able to carry out planning and follow-up activities
Katz and Kahn (1978)	Project manager should have technical skill, human skill, and conceptual skill
Stuckenbruck (1976)	A proficient manager must be: Multi-disciplinarily oriented; Global problem oriented i.e. he must consider the external, political, legal, and environmental aspects; Effective problem solver and decision maker; A good manager and administrator; he should master the basics of the management of planning, budgets, supervision, and follow-up; Possessing good analytical abilities; Creative in dealing with information and problems; An effective communicator; Motivating his team members to achieve fixed goals; Flexible, and able to adapt to change; Having the right temperament, and should be able to keep his calm, and should be realistic, dedicated, generous, stable, quick thinking, disciplined and persistent
Adams and Barndt (1978)	Project manager must also be able to work on planning, coordination and budgeting, the technical assessment of financial reports, and on the customer-salesman relationship in addition to managing his team
Fryer (1979) cited in Odusami	Managing change, recognizing opportunities, handling problems, decision making, and social skills are the skills needed in a project manager
Spitz (1982) cited in Pettersen (1991)	Inter personal skill- defined as the ability to communicate effectively Skills for synchronizing different technology–the ability to put the different fractions of the project into order Expertise- involves technical knowledge connected with the product, process or market covered by the project Information processing skill- this allows the manager to obtain, use and disseminate information
Goodwin (1993)	Project manager's effectiveness will depend on conceptual, human, and negotiating skills as well as, to a lesser extent, on technical skills. The ingredient that is common to the range of skills necessary to his or her effectiveness is the ability to communicate both verbally and in writing
Meredith et al. cited in El-Sabaa (2001)	The skills needed for a project manager are categorized into six-skill area: communication, organizational, team building, leadership, coping, and technological skills

(continued)

Table 8.2 (continued)

Authors	Skill description
El-Sabaa (2001)	He has found the three major skill groups. These are Human skills, Conceptual skills, and Technical skill. The different skills falling under these broad skill groups are given below Human skills: This includes mobilization, communication, coping with situations, delegation of authority, political sensitivity, high self-esteem, and enthusiasm Conceptual skills: This includes skills of planning, organizing, having strong goal orientation, ability to see the project as a whole, ability to visualize the relationship of the individual project to the industry and the community and strong problem orientation Technical skill: This includes specialized knowledge in the use of tools and techniques, project knowledge, understanding methods, process and procedures, the technology required, and skill in the use of computer
Project management body of knowledge (2001)	The project manager should have the following characteristics: Attitude–an open positive "can do" attitude, which encourages communication and motivation, and fosters co-operation Common sense—a strong ability to spot sensible, effective, straight forward, least risky, least complex solutions i.e. 90 % right on time is better than 100 % far too late! Open mindedness—an approach where one is always open to new ideas, practices and methods and in particular gives equal weight to the various professional disciplines involved on the project Adaptability—a propensity to be flexible where necessary and avoid rigid patterns of thinking or behavior, to adapt to the requirements of the project, the needs of the sponsors, its environment and people working on it and for it to ensure a successful outcome Inventiveness—an ability to discover innovative strategies and solutions either from within oneself or by encouragement with other members of the project team and to identify ways of working with disparate resources to achieve the project objectives Prudent risk taker—a willingness and ability to identify and understand risks but not to take a risky approach in an unwise or reckless fashion Fairness—a fair and open attitude, which respects all human values Commitment—a very strong over-riding commitment to the project's success, user satisfaction and team
Kerzner (2002)	Team building, Leadership, Conflict resolution, Technical expertise, Planning, Organization, Entrepreneurship, Administration, Management support, and Resource allocation

mechanical, plant and machinery, HVAC, accounts, materials, design, construction method, quality, safety, and HRD totaling to 12 numbers to say the least. In case the project coordinator is not there, the project manager has to coordinate with above-mentioned 12 departments. It can be seen that with these 12 function lines, the possible coordination routes may be $12 \times (12-1) \leq 132$. It can be imagined the amount of difficulty the designated project manager would have in coordinating resources for his sites. This will leave him completely drained out and the project manager would not be able to attend to other important project requirements. It is in these circumstances that role of a project coordinator is considered vital.

In terms of hierarchy, Kerzner (2002) places project coordinator in between project administrator and technical assistants and finds planning, coordinating, analyzing, and understanding of organization as the required skill to carry out his responsibility. In the subsequent sections, the skill requirement of a project coordinator has been examined. As an initial step in this process, previous research results on project manager's skill requirement have been consolidated which were then modified after pilot survey and interviews with professionals (Songer and Molenaar 1997). These interviews resulted in a final list of 24 traits. For the better comprehension of the reader the 24 traits are explained in Table 8.3 (Katz 1974; Pettersen 1991; Goodwin 1993; Kerzner 2002; El-Sabaa 2001).

The above traits are then used in the first stage questionnaire. Respondents are asked to evaluate the traits or skills possessed by the project coordinators of both successful and failed projects, on a 1 to 5 scale. It is hypothesized that the traits of project managers of successful projects are different from that of failure projects. Therefore the mean values of responses on traits of successful projects are compared with that of failure project with null hypothesis,

H_0: $\mu_1 = \mu_2$ and alternate hypothesis, H_1: $\mu_1 \neq \mu_2$ where μ_1 and μ_2 are the mean values of responses on individual trait of successful and failure project respectively. Significance level of 5 % is considered for the hypothesis testing. Subsequently it is also tried to evaluate the extent of dissimilarity between these mean values in individual cases.

It is found that statistically significant difference exists between each trait of a successful project's coordinator and that of a failed project's coordinator. This proves that the skill set possessed by the project coordinator is also an important factor that directs the outcome of any construction project. The next section analyses these traits in detail.

8.5 Dominant Traits of Coordinators in Successful and Failure Projects

In order to distinguish those personal characteristics or traits that are dominant in the successful projects' coordinators from that of failure projects' coordinators, relative importance of the 24 traits are found based on the mean values of the

Table 8.3 Glossaries of project coordinator's traits (Jha and Iyer 2006, with permission from Taylor and Francis)

Description of traits	Definition
Timeliness	Ability to successfully manage multiple tasks within given time constraints
Maintaining records	Skill of keeping a diary and keeping notes
Interpersonal skill	Skill to mix in friendly converse
Relationship with client, consultant and contractor	Skill in maintaining good human relations with client, consultant, and other contractors
Technical knowledge of the subject	The capacity to manage the technological innovation and integration of solutions for the success of the project. Understanding of complex elements required to effectively complete tasks associated with a given profession
Coordination for achieving quality	Ability to manage production of goods or services within a clearly defined set of expectations
Liaison skill	Ability to channelize communication between groups
Knowledge of project finance	Ability to understand financial statements and financial ratios, and to deal with accounting firms and financial institutions
Communication skill	Ability to interact effectively with others at all levels within and outside the organization
Reliance on systematic approach	Skill to do things methodically and not in a haphazard manner; A series of orderly action at regular hours
Understanding of contract clauses	The power to understand, the capacity for rational thought of contract clauses
Monitoring skills	Ability to observe something (and sometimes keeping a record of it), showing quick and keen perception
Planning skills	This involves the preparation of a project summary plan before the project starts and requires communication and information processing skills
Forecasting skills	Skill of predicting or foretelling about the future by looking at the present status
Facilitating skills	Skill to make easy or less difficult, the execution of a task
Resource utilization skills	The program manager needs to work out specific agreements with all key contributors and their superiors on the tasks to be performed and the associated budgets and schedules
Belief in team playing spirit	The ability to integrate people from many disciplines into an effective team
Analytical skills	Ability to look logically at a technical situation
Concern for other's ego	Not to remain self-centered and respecting other's individuality; Regard for other's interest, power and happiness
Concern for conciliation	The act of placating and overcoming distrust and animosity
Motivating skills	Ability to influence others to contribute to attaining firm's goals

(continued)

Table 8.3 (continued)

Description of traits	Definition
Follow up quality	Pursuance or skill for the continuance of something begun with a view to its completion
Concern for safety, health, and welfare of labor and employees	Interest or feeling for safety, health, and welfare of labor and employees
Understanding of human psychology	Understanding the science of the human soul, specifically the systematic or scientific knowledge of the powers and function of the human soul

responses on individual traits. As can be seen from the scale structure of the question Q7, lower is the mean value of the trait, higher the extent to which this particular trait is possessed by a project coordinator. The mean values of responses are summarized in Table 8.4.

The most dominant skill possessed by the project coordinators of successful projects are *relationship with client, consultant and contractor* and this has the lowest mean value of 2.00. This is followed by *timeliness, and technical knowledge of the subject* both with a mean score of 2.04. *Belief in team playing spirit* (2.07), and *Coordination for achieving quality* (2.08) have been ranked at number 4 and 5 respectively. The five least dominant traits out of a total of 24 traits are *knowledge of project finance* (2.47); *understanding of human psychology* (2.50); *concern for conciliation* (2.54); facilitating *skills* (2.59); and *concern for other's ego* (2.77). As the mean score for these traits would suggest the traits possessed by the coordinators are still having good to very good values.

For failed projects; the top ranking traits (rank one to five) of the project coordinator: *technical knowledge of the subject; understanding of contract clauses; maintaining records; concern for safety, health, and welfare of labor and employees;* and *coordination for achieving quality* shows that project coordinators did possess high quality of these traits, and yet the projects have failed. It only indicates that project manager has given importance to these activities and might have neglected some other vital characteristics. This can be seen from high mean values and the low rank of a few traits such as: *relationship with client, consultant and contractor* (rank 13); maintaining *timeliness* (rank 18); and *belief in team playing spirit* (rank 20). The coordinators for failed projects also lack in *motivational skill* (rank 19). These low ranked traits explain generally the poor human relationship of the project coordinator. In words of Katz (1974) it can be summarized that project coordinators for failure projects lack human relationships skill. A project coordinator has to interact with number of people who may not be under his direct control and for performing his duties he has to take help invariably from his colleagues and superiors and the human relationships skill becomes of too much importance for the coordinators.

From Table 8.4, it can be observed that for some traits the difference in rank is large, which indicates that the particular trait is dominant in one group and less noticeable in other group. For example, *relationship with client, consultant, and*

Table 8.4 Comparison of rank of coordinator's traits for successful and failed projects (Jha and Iyer 2006, with permission from Taylor and Francis)

S. No.	Skill description	Mean of successful projects	Rank of successful projects	Mean of failed projects	Rank of failed projects
1	Relationship with client, consultant and contractor	2.00	1	3.17	13
2	Timeliness	2.04	2	3.33	18
3	Technical knowledge of the subject	2.04	2	2.65	1
4	Belief in team playing spirit	2.07	4	3.36	20
5	Coordination for achieving quality	2.08	5	2.90	5
6	Understanding of contract clauses	2.11	6	2.80	2
7	Monitoring skills	2.12	7	3.12	11
8	Maintaining records	2.13	8	2.81	3
9	Planning skills	2.17	9	2.93	6
10	Liaison skill	2.18	10	3.10	10
11	Follow up quality	2.18	11	3.00	7
12	Reliance on systematic approach	2.20	12	3.00	7
13	Motivating skills	2.22	13	3.33	19
14	Communication skill	2.29	14	3.21	15
15	Interpersonal skill	2.30	15	3.12	12
16	Resource utilization skills	2.32	16	3.26	17
17	Concern for safety, health, and welfare of labor and employees	2.37	17	2.88	4
18	Analytical skills	2.40	18	3.17	13
19	Forecasting skills	2.46	19	3.38	22
20	Knowledge of project finance	2.47	20	3.23	16
21	Understanding of human psychology	2.50	21	3.05	9
22	Concern for conciliation	2.54	22	3.37	21
23	Facilitating skills	2.59	23	3.45	24
24	Concern for other's ego	2.77	24	3.40	23

contractor (rank 1 and 13 in successful and failed projects respectively); *timeliness* (rank 2 and 18 in successful and failed projects respectively), *belief in team playing spirit* (rank 4 and 20 and in successful and failed projects respectively) etc. The dominance can also be seen from the cumulative frequency chart drawn for some of these attributes for illustration purpose in Fig. 8.1. The concavity downward on the cumulative frequency as against the concavity upward (i.e., convexity) shows the dominance of the attribute possessed by the coordinator in successful projects when compared to failed projects.

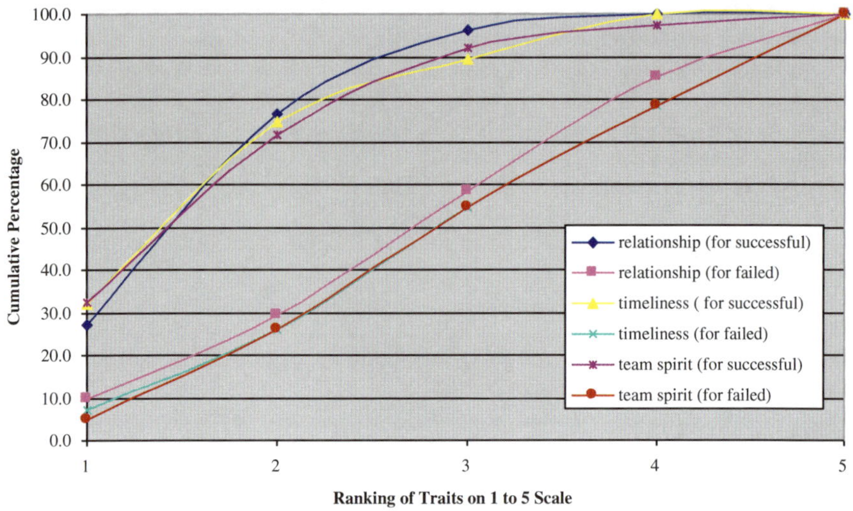

Fig. 8.1 Cumulative frequency chart of coordinator's skill (Jha and Iyer 2006, with permission from Taylor and Francis)

8.6 Major Trait/Skill Category

Since traits are mostly composed of psychological or behavioral or appropriate knowledge aspects and they cannot be viewed independent of each other and any change in one will have automatic effect on many other variables. In the present case too it can be observed that most of the skills are correlated with each other. Hence factor analysis is considered appropriate here for analyzing the responses of Q7 and then to understand them in groups of concomitant variables. The factor analysis of the responses on project coordinators' traits of successful projects have resulted into three major skill categories and explained a total variance of 67.44 % while the factor analysis of responses on failed projects did not yield significant and meaningful results. The factor description of skills of coordinator of successful projects along with variance explained by each factor is given in Table 8.5.

8.6.1 Team Building Skill

It is the first factor explaining the maximum variance of 26.86 % out of 67.44 % of total variances explained. It can be observed from any construction project site, a project coordinator has to carry out his work within limited authority. Unless his team members have confidence in him things are not likely to work for the coordinator. Traits emerging under this skill group encompass the human relationships skill suggested by Katz (1974) for the project manager's trait. Human

Table 8.5 Factor profile of project coordinator's attributes (Jha and Iyer 2006, with permission from Taylor and Francis)

Details of factor and the attributes	Factor loading
Factor_1 Team Building Skill (variance explained 26.86 %)	
Concern for conciliation	0.805
Concern for other's ego	0.751
Understanding of human psychology	0.710
Analytical skills	0.708
Motivating skills	0.675
Belief in team playing spirit	0.671
Timeliness	0.608
Facilitating skills	0.604
Interpersonal skill	0.595
Communication skill	0.575
Technical knowledge of the subject	0.565
Resource utilization skills	0.564
Factor_2 Contract Implementation Skill (variance explained 21.56 %)	
Reliance on systematic approach	0.768
Understanding of contract clauses	0.723
Concern for safety, health, and welfare of labor and employees	0.719
Monitoring skills	0.698
Maintaining records	0.564
Follow up quality	0.557
Forecasting skills	0.503
Planning skills	0.509
Factor_3 Project Organization Skill (variance explained 19.01 %)	
Relationship with client, consultant, and contractor	0.724
Coordination for achieving quality	0.698
Knowledge of project finance	0.658
Liaison skill	0.584
Planning skills	0.515
Monitoring skills	0.504
Timeliness	0.530
Interpersonal skill	0.534
Communication skill	0.500

relationships skill involves the ability to communicate effectively, and maintain a harmonious working group. The ability to motivate employees also falls in human relationships skill as suggested by Katz. Team building requires conciliatory approach and not the confrontationist approach. A coordinator needs to show concern for other's ego and must have a sound understanding of human psychology. Most importantly a coordinator must believe in the team spirit. A coordinator must be able to communicate properly, both through verbal and written communication, and he must be proficient in interpersonal skill. It is to be kept in mind that during the course of fulfillment of his duties, a coordinator has to interact with different departments, which may not be under his direct control and under such situation he must possess team building skill and project himself as a team member. Team building skill has been defined by Kerzner (2002) as the ability to integrate people from many disciplines into an effective team and he finds that team building as one of the essential skills for program manager.

8.6.2 Contract Implementation Skill

This is the second factor explaining 21.56 % out of the total variance 67.44 %. Contract implementation is one of the major groups that emerged from the factor analysis of the twenty important coordination activities performed by a project coordinator to achieve day-to-day coordination (Chap. 7). A coordinator is supposed to assist the project manager fulfilling the contractual promises. The reliance on systematic approach and a sound understanding of contract clauses make a project coordinator understand his responsibilities toward fulfilling this duty. A project coordinator with monitoring and forecasting skill can keep a close watch on schedule and cost of the project and appraise the PM of any deviation from the same. Subsequently with his follow up skill he can push his team members to correct the deviations to bring the project back on time and cost requirement. Safety, health and welfare of employees are one of the important contract requirements and project coordinator's concern for the same cannot be underestimated. Maintaining records of all the important events is also an important function and it helps in reducing the disputes at a later date.

8.6.3 Project Organization Skill

This is the third factor explaining 19.01 % variance. This group of traits suggests that the project coordinator must be able to perceive the project as a system by keeping a global perspective and not thinking one aspect at a time. The project coordinator must be good at keeping good working relationship with client and consultants. A project coordinator has to work with many different groups or departments to perform his duties, and for this he needs to ensure cooperative relationships. To achieve proper relationship he must be good at interpersonal skills and should have good communication skills. The project coordinator must be good at liaison and he should also ensure proper quality of workmanship. The project coordinator must be able to plan, he should have requisite knowledge of project finance and he should be able to ensure timeliness. The importance of possessing proper communication and interpersonal skill has already been dealt with under the first group of skills.

The challenge facing a future construction project coordinator is the development and successful application of these important skills to achieve their project objectives. Some of these skills can be acquired in schools, while others may be acquired on the field. Kerzner (2002) suggests experiencial learning, on the job training, formal education and special courses, professional activities, seminars and readings as some of the ways to train a project manager. There is no reason to believe that these ways won't work for training a project coordinator. The ranking of the importance of skills can be a guide in the training of construction project coordinators both at the undergraduate or postgraduate levels and post qualification. They can also be used as a yardstick in appointing/selecting a future construction project coordinator during interview and final selection.

8.7 Summary and Conclusions

In this chapter, 23 elements affecting coordination have been identified and are ranked based on their mean scores of responses on these elements on two parameters: *intrinsic characteristic* representing its criticality and *increase in coordination efforts* due to these elements. Further, 24 traits of the key person of the project, the project coordinator are identified through literature and personal interviews. Importance of these traits is assessed through analysis of responses on these traits. Important conclusions drawn from analyses are given below.

- While 22 out of 23 coordination elements are found to be critical and 18 elements cause increase in coordination efforts of varying degrees and project coordinators require to be attentive in handling these elements, project coordinators are required to be more alert in two highly critical elements: *nonavailability of finance for the project;* and *very high degree of hazard associated with the project.*
- There exists significant difference between each trait of a successful project's coordinator and that of a failed project's coordinator.
- The traits of project coordinators are not the independent entity, but they are correlated with each other and show concomitant variation with each other. These elements together represent three latent properties: *team building skill; contract implementation skill;* and *project organization skill.*

References

Adams, J. R., & Barndt, S. E. (1978). Organizational life cycle implication for major projects. *Project Management Quarterly, 9*(4), 32–39.

El-Sabaa, S. (2001). The skills and career path of an effective project manager. *International Journal of Project Management, 19*, 1–7.

Forsberg, K., Mooz, H., & Cotterman, H. (1996). *Visualizing project management*. New York: Wiley.

Fryer, B. (1979). *Management development in the construction industry* (pp. 16–18). May: Building Technology and Management.
Gaddis, P.O. (1959). The project manager. Harvard Business Review. May/Jun, 89–97.
Goodwin, R. S. C. (1993). Skills required of effective project managers. *Journal of Management in Engineering, ASCE, 9*(3), 217–226.
Iyer, K.C., and Jha, K.N. (2004). Factors affecting coordination in Indian construction projects. Proceedings of international symposium on project procurement for infrastructure systems (CIB W92), IIT Madras, Chennai (pp. 283–291), Jan 7–10.
Jha, K. N.,& Iyer, K. C. (2006). Critical determinants of project coordination, *International Journal of Project Management, 24*(4), 314–322.
Katz, R.L. (1974). Skills of an effective administrator. Harvard Business Review, Sep/Oct, 90–102.
Katz, D., & Kahn, R. L. (1978). *The social psychology of organisations*. USA: Wiley.
Kerzner, H. (2002). *Project management: A systems approach to planning, scheduling, and controlling*. New Delhi: CBS Publishers & Distributors.
Laufer, A., Woodward, H., & Howell, G. A. (1999). Managing the decision-making process during project planning. *Journal of Management in Engineering. ASCE, 15*(2), 79–84.
Odusami, K. T. (2002). Perceptions of construction professionals concerning important skills of effective project leaders. *Journal of Management in Engineering. ASCE, 18*(2), 61–67.
Pettersen, N. (1991). What do we know about the effective project manager? *International Journal of Project Management, 9*(2), 99–104.
PMBOK (2001). A guide to the project management body of knowledge. Project management institute. http://www.pmi.org.
Songer, A. D., & Molenaar, K. R. (1997). Project characteristics for successful public-sector design-build. *Journal of Construction Engineering and Management. ASCE, 123*(1), 34–40.
Spitz, C. J. (1982). *The project leader: A study of task requirements, management skills and personal style*. USA: Doctoral Dissertation Case Western Reserve University.
Stuckenbruck, L.C. (1976). The ten attributes of the proficient project manager. Proceedings of the 8th Project Management Institute Seminar/Symposium, Montreal, Canada.
Tarricone, P. (1992). Portrait of a manager. *Civil Engineering—ASCE, 62*(8), 52–54.
www.tbs-sct.gc.ca

Chapter 9
Summary and Conclusions

Abstract Based on project management literature, case studies and experts interviews 55 project performance attributes were identified. Responses were sought through a two-stage questionnaire survey to identify the critical success factor for four project performance criteria: schedule, cost, quality, and no-dispute. Critical success factors vary across the four performance criteria and the impact they have on project performance is dependent on the current project performance level. Some of the critical success factors have been used as predictor variables to predict the project performance using a neural network model. Structural Equation Modeling has shown that among the success factors, 'human factors' and 'management actions' have positive interrelationship with the project success. Coordination among project participants is also found to be important for project success. Various coordination issues have been discussed and the required traits of a project coordinator have been specified.

9.1 Summary of Work Done

In the present study the following broad objectives have been addressed.

1. To identify critical factors responsible for success or failure of projects.
2. To evaluate the relative impact of the above factors on the performance of a project.
3. To identify the predictor variables used for predicting the schedule, cost, quality, and dispute performance of a construction project and to develop a user interactive model to predict the performance of a construction project.
4. To examine the hypothesis that 'project success' is influenced by 'success traits' and to explore the impact of the success traits on project success.
5. To evaluate impact of coordination and coordination related activities on success of projects.

K. N. Jha, *Determinants of Construction Project Success in India*,
Topics in Safety, Risk, Reliability, and Quality 23, DOI: 10.1007/978-94-007-6256-5_9,

6. To study issues involved in coordination and to find out the required traits of a project coordinator.

The research method to achieve the above objectives included: identification of project performance evaluation criteria; project attributes responsible for success/failure of a performance; coordination activities required for the success of a project; and other allied elements affecting coordination through literature survey and personal interviews with professionals. Due to non-availability of documented/structured data of completed projects, questionnaire survey approach was considered for data collection. The questionnaire survey was conducted in two stages: first stage questionnaire which was general in nature and the second stage questionnaire which was specific. A total of 114 responses were received out of 400 mailed from the professionals having long and rich experience for the first stage questionnaire and 92 responses were received out of 300 questionnaires sent for the second stage questionnaire. There were 13 questions in the first stage questionnaire which sought information on the following.

- Respondent's personal and organization details;
- The details of one recently completed project by the respondents;
- Relative importance of different project stages in respondent's opinion;
- The respondent's views on priority of performance criteria;
- The respondent's views on the criticality of project performance attributes;
- The relative importance of coordination activities; and
- The criticality and influence of key elements on the coordination effort.

The second stage questionnaire contained eight broad questions seeking information on the following.

- Project details of the choice project (here choice project was the one which the respondent would be familiar with);
- Performance level (intensity of success or failure) of choice project;
- Extent of contribution of the critical success and failure factors in the choice project; and
- Extent to which important coordination activities have been performed on the choice project.

The responses for the first stage questionnaire were analyzed first using statistical software SPSS release 9. The statistical tests in this study have included both univariate and multivariate analysis techniques. Univariate analyses included finding out summary statistics of responses such as means, standard deviations and frequencies; hypotheses testing using t-test; one-way analysis of variance (ANOVA); and nonparametric Spearman's rank correlation technique. Multivariate analyses in the study comprised of two techniques: factor analysis and multiple regression.

9.2 Identification of Critical Factors Responsible for Success and Failure of Projects

Based on the mean responses, the project attributes were classified distinctly in two categories: *success attributes* those which contribute for the success of a project and *failure attributes* which contribute negatively for the outcome of a project.

Responses on the success attributes and failure attributes were then subjected to factor analysis separately on the four performance criteria for better understanding of their inherent properties. Since factors extracted under different performance evaluation parameters (schedule, cost, quality, and no-dispute) are different, a union of all factors were considered for further study. A total of 20 factors (consisting of 11 success factors and 9 failure factors) were formed from the analyses on the four performance criteria and they were taken for further study through the second stage questionnaire survey. These factors are given in Table 9.1.

9.3 Evaluation of Relative Impact of Success and Failure Factors on Performance of Project

To achieve the next objective of the study, i.e., finding the relative impact of above critical success and failure factors on project performance the second stage questionnaire survey was conducted as discussed previously. In this questionnaire,

Table 9.1 Summary of success and failure factors

S. No.	Success factors	S. No.	Failure factors
1.	Project manager's competence	1.	Conflict among project participant
2.	Top management support	2.	PM's ignorance and lack of knowledge
3.	Monitoring and feedback by project participants	3.	Hostile socio economic environment
4.	Favorable working condition	4.	Owner's incompetence
5.	Commitment of all project participants	5.	Indecisiveness of project participants
6.	Owners competence	6.	Harsh climatic condition at site
7.	Interaction between project participants-internal	7.	Aggressive competition during tendering
8.	Interaction between project participants-external	8.	Negative attitude of project participants
9.	Good coordination among project participants	9.	Faulty project conceptualization
10.	Availability of trained resources		
11.	Regular budget update		

the factors summarized in Table 9.1 were used as variables and responses were sought on the extent of contribution of these factors in the performance of the choice project. Through a separate question in the second stage questionnaire the actual performance of the choice project was sought. Data points (responses of the questionnaire) being discrete in nature, multinomial logistic regression was used in which the performance rating of the choice project was taken as predictor variable and extent of contribution of various factors were considered as criterion variables. Important findings of the analyses are given below:

- Fourteen factors out of the 20 success and failure factors identified in the study have been found to be significant under different performance criteria.
- Seven factors are observed to have significant influence on the schedule outcome of a project. While the factors: *commitment of the project participants*; *owner's competence*; and *conflict among project participants* have been found to possess the capability to enhance performance level, the remaining four factors: *coordination among project participants*; *project manager's ignorance and lack of knowledge*; *hostile socio economic environment*; and *indecisiveness of project participants* tend to retain the schedule performance at its existing level.
- Factors *coordination among project participants*; *favorable working condition*; and *conflict among project participants* are found to be important factors to enhance cost performance of a project. On the other hand, important factors like *top management support*; *commitment of project participants*; *project manager's ignorance*; and *indecisiveness of project participants* tend to keep the cost performance of a project at the same level.
- Five success factors have significant influence on the quality performance. Out of these, the three factors: *project manager's competence*; *top management support*; and *interaction between project participants-external* contribute significantly in enhancing the project quality performance from its existing level, the remaining two factors: *owner's competence*; and *interaction between project participants-internal*, tend to retain the quality performance at the existing level itself. Emergence of project manager's competence and top management support as positive contributor to improve quality reestablishes the findings of quality gurus that management is more responsible to achieve the desired quality in any system.
- Six factors have emerged to be significant corresponding to no-dispute criteria. Out of these the factors: *top management support*; *favorable working condition*; and *owner's competence* contribute significantly in avoiding disputes. These factors enhance the probability to avoid dispute only when performance on no-dispute rating is of average nature. None of the success factors considered in the present study seems to have dispute avoiding potential either at low or high dispute ratings.
- Contradictory to the common belief, the factor *conflict among project participants* is observed to contribute in improvement of the project performance, and this is also in line with the conclusions of some of the past researchers and if it is exploited appropriately it will fetch better results.

- While impact of various factors on project performances has been evaluated in the present study, methodology to measure and alter the current level of individual factors is yet to be explored. However, the findings given about are expected to give a broad guideline to any professional to select appropriate factor for enhancement or sustenance of the desired level of performance.

9.4 Project Performance Prediction

The most important predictor variables for a project performance have been identified from correlation analysis and a neural network based project performance prediction model has been developed.

- For schedule performance prediction model, the predictor variables are: *project manager's competence, monitoring and feedback by project participants, commitment of all project participants, owner's competence, interaction between external project participants,* and *good coordination between project participants.* The ANN models have a feed-forward network based on a back-propagation algorithm, in which the 6-3-1 structure has given the least MAPD of 11 %. The high degree of predictive ability shows that the predictor factors are correct and can be used to predict the schedule performance of the construction project.
- For the cost performance prediction model, the predictor variables are: *project manager's competence, commitment of all project participants, owner's competence,* and *good coordination between project participants.* The best ANN model had a 4-4-1 feed forward network structure based on back propagation algorithm and gave the least MAPD of 10.379 %. The high degree of predictive ability show that the factors identified from correlation analysis are correct and can be used to predict the cost performance of the project.
- For the quality performance prediction model, the predictor variables are: *project manager's competence*, *monitoring and feedback by project participants*, *commitment of all project participants*, *good coordination* between *project participants*, and *availability of trained resources*. The developed ANN models had a feed forward network based on back propagation algorithm and the 5-5-1 structure gave the least MAPD of 8.044 %. The high degree of predictive ability show that the factors identified can be used to predict the quality performance of the construction project.
- For the no-dispute performance prediction model the factors such as: monitoring and feedback by project participants, favorable working conditions, owner's competence, good coordination between project participants and regular budget update are significant. The 5-3-1 feed forward neural network structure based on Levenberg–Marquardt algorithm and tangent sigmoid transfer function gave the least MAPD of 10.107 %.

- A user-interactive model to predict the project performance of a construction project was illustrated that provides simple access to the developed ANN model and subsequently automates performance prediction.

9.5 Success Traits for a Construction Project

Out of the eleven success factors derived earlier, 'human factors' and 'management actions' have been further analyzed. The hypothesized positive interrelationship between success traits and project success tested using structural equation modeling technique has been found to hold good. It is thus concluded that trained, committed, competent participant's coordination, with constant monitoring and feedback with regular budget update will influence the successful completion of a project.

9.6 Evaluation of Impact of Coordination and Coordination Related Activities on Success of Projects

As stated in the beginning of the section, through the literature survey and personal interviews 59 coordination activities were identified. The key elements affecting coordination in a construction project and traits of a good coordinator were also identified and evaluated based on the responses of the first stage questionnaire. Based on the mean responses on 59 coordination activities the following 20 coordination activities were identified to be important.

- Implementing all contractual commitments
- Arranging timely carrying out of all tests for inspections and approval by the engineer and maintaining records of the same
- Arranging submission of samples of materials for approval by the engineer
- Application of good technical practices
- Preparation of a project quality plan in line with contract specification
- Arranging remedial work methods and programs for executing in case of defect or damage
- Identification of appropriate human resources, materials, and equipments for the project
- Estimation of the optimum resource requirements
- Proper assignment of task to the available human resources for the project
- Organization of resources (manpower, plant, and material) for effective utilization

- Ensuring discipline among all employees
- Resolving differences/conflicts/confusion among participants
- Motivation of project participants
- Development of a team spirit and receiving constructive input from all participants in the project
- Identification of activities on critical path
- Regular monitoring of critical path activities for adhering to schedule
- Arrangement of required inputs like drawings, specifications, and technical details on time for execution
- Agreement on detailed methods of construction with all the parties involved
- Analysis of the project performances on time, cost and quality, detecting variances
- Monitoring the overall functioning of each section and department of the project.

These 20 coordination activities were used subsequently in the second stage questionnaire and responses were sought on the extent of achievement of these activities on the choice projects. The responses were subjected to factor analysis which identified the activities in four broad categories of managerial functions—*Planning functions; Resource handling functions; Contract implementation functions*; and *Team building functions*.

Subsequent to identification of coordination activities as four distinct managerial functions (factors), stepwise multiple regression was carried out using the factors as explanatory variables and the extent of contribution of coordination in the outcome of the choice project (a separate question in the second stage questionnaire) as response variable. This has identified that *resource handling function* is the most critical of all functions in achieving good coordination.

- Relative importance of various coordination activities vary depending on the performance measure employed to measure the success. For example, when schedule criteria is given the supreme priority, the most important activity for a coordinator is 'regular monitoring of critical path activities for adhering to schedule' whereas when cost is taken as the prime performance criteria 'monitoring the budget on all activities and taking corrective action' takes supreme importance. Similarly 'application of good technical practices' and 'implementation of all contractual commitments' turn out to be the most important coordination activities for quality and no-dispute performance criteria respectively.
- The study identifies that six coordination activities are significantly important in influencing the coordination rating of a project. These activities are *preparation of a project quality plan in line with contract specification*; *arranging remedial work methods and programs for executing in case of defect or damage*; *estimation of the optimum resource requirements*; *development of a team spirit and receiving constructive input from all participants in the project*; *arrangement of required inputs like drawings, specifications, and technical details on time for execution*; and *agreement on detailed methods of construction with all the*

parties. Most important of them are *estimation of the optimum resource requirements* and *agreement on detailed methods of construction with all the parties*. They show their potential of continual improvement of coordination levels if handled properly toward enhancement.

- Project coordination is not an isolated and independent activity, but is a typical management function having its inherent role of varying degrees in all the major management activities that are broadly represented by the four factors, i.e., *Planning*, *Resource handling*, *Contract implementation*, and *Team building*. The most critical factor among them has been identified to be 'Resource handling' indicating that if this function is handled properly major coordination can be considered to have been achieved for the project.

9.7 Project Coordination for Success

On the allied issues related to coordination, the mean values of responses on the key elements affecting coordination provided the relative importance of the key elements. These elements when experienced by a project manager in his project site, he may have to be extra careful.

Similarly, on another allied issue of coordination, the important traits of coordinators of successful projects are compared with that of failure projects which showed a significant difference between the traits of two types of coordinators. Factor analysis of responses on traits of project coordinators of successful projects yielded that the traits could be classified in three broad types of skills: *team building skill; contract implementation skill*; and *project organization skill*. The factor analysis of responses on traits of project coordinator of failed projects did not give any meaningful result.

- Among the related issues to coordination while 22 out of 23 coordination elements are found to be critical and 18 elements cause increase in coordination efforts of varying degrees and project coordinators are required to be attentive in handling these elements, project coordinators are required to be more alert in two highly critical elements: *nonavailability of finance for the project*; and *very high degree of hazard associated with the project*.
- There exists significant difference between each trait of a successful project's coordinator and that of a failed project's coordinator.
- The traits of project coordinators are not the independent entity, but they are correlated with each other and show concomitant variation with each other. These elements together represent three latent properties: *team building skill; contract implementation skill;* and *project organization skill*. While selecting a project coordinator for a given job, presence of these qualities in a coordinator may be ensured by the top management to get expected performance.

Finally the conclusions drawn from the study are validated through case study and structured interviews with professionals. Important conclusions of the study are given in the next section.

9.8 Limitations

As with any other opinion-based study, the present study also has some limitations or weaknesses. The weaknesses identified by Morledge and Owen (1999) associated with the application of CSF are pertinent for the present study as well. These weaknesses are: Subjectivity; Bias; Human inability to process complex information; Time dependency of variables; Imprecise definitions, generalization; and Qualitative performance measures. In addition to these weaknesses some other limitations related to the present study are given below:

- The majority of respondents have evaluated the choice project in their execution stage only and very few have evaluated the performance of choice projects in planning and operation stages. Hence the study has a limitation in this sense.
- The majority of the project presented by the respondents belongs to building projects and hence the model derived will be more suited for building projects.
- In the present study the overall success has been assumed to be the weighted average of success in schedule, cost, quality, and no-dispute individually.
- The regression model described in the study does not pay attention to the long-term success of the construction project.
- The regression model adopted in the study has been constructed from the inputs mainly received for successful projects. Hence the study would apply to those projects, which had met their objectives successfully.

9.9 Suggestions for Further Studies

- The study has identified the critical factors corresponding to four performance objectives: schedule, cost, quality, and no-dispute. Similar studies can be undertaken to identify the critical factors corresponding to other project objectives e.g. for safety compliance, technical innovation, and participant satisfaction.
- Further studies can also be undertaken to develop project performance prediction models using the critical factors. These models can be developed using either mathematical tool like neural network or statistical tool like regression models. Data on critical factors can be collected from project.

- The present study has identified success factors corresponding to execution phase of a project. More such studies should be conducted to explore the critical factors corresponding to other project phases such as planning phase, and operation phase.
- The measurement technique for different critical success and failure factors should be developed. Similarly evaluation tool for measuring the level of key coordination activities should also be undertaken.
- The comparative study of cost aspect involved in enhancing the different critical success factors and reducing the critical failure factors need to be undertaken.
- Framework for performance assessment for a project coordinator can be developed using the three performance variables: *team-building skill, contract implementation skill,* and *project organization skill* obtained from the study.

Reference

Morledge, R., & Owen, K. (1999). Developing a methodological approach to the identification of factors critical to success in privatized infrastructure projects in the UK. In: S. Ogunlana (Ed.), *Proceedings of Profitable Partnering in Construction Procurement, CIB W92* (pp. 487–498). Publication 224.

Index

K. N. Jha, *Determinants of Construction Project Success in India*,
Topics in Safety, Risk, Reliability, and Quality 23, DOI: 10.1007/978-94-007-6256-5,

O

P

Q

R

Printed by Printforce, the Netherlands